The Savvy Scientist

John Corthell, PhD

Frugal Ink

Disclaimers

- (advice, how-to)

Cover Illustration Copyright © 2020

Copyright

Dedication and Thanks

To my wife, who knows all of my faults but likes me anyway

To my Ls and Js, just because I love you

To all the savvy scientists out there, new and old

Table of Contents

Introduction

"Research is four things: brains with which to think, eyes with which to see, machines with which to measure, and fourth, money." -Albert Szent-Gyorgyi

As a graduate student or new researcher, you have two main resources: money and time.

You may not personally have a lot of money, but your training lab might. Many of the experiments you plan out and run will involve the balancing act of money: you can pay money to get more time (i.e., buy a solution ready-made instead of making it yourself) or use your time to keep more money (i.e., make the solution yourself since that money can be used for other things).

You will need to pay attention to money and get as much value out of each dollar as you can to fund your experiments, fund your lab, fund additional people (the greatest lab expense), and save up for that fancy machine you would really like to use.[1]

The rapid expansion of the NIH budget in the late 1990's to 2003 allowed many scientists to secure grant funding and allowed many more labs and colleges to hire and expand than could previously...but it's not the 1990s anymore.

I started going to graduate school when that funding supply was rapidly dwindling between all the hands that needed a piece of that pie. I was also in a lab where I had a fixed budget, so I needed to

---

[1] You know, the combination flow cytometer-espresso maker. Caffeinated cells express green fluorescent protein (GFP) better, right?

get grants if I wanted to get my PhD; however, when I started grad school, I had few opportunities for grant funding and too little experience to get started.

So I learned to run my experiments cheaply while learning to run them effectively.

While this was happening, life was happening at a frantic (for grad school) pace. I broke two of the cardinal rules of grad school:

#1. I got married while going to graduate school.[2]

#2. We had children (plural) while going to graduate school.[3]

And so I needed to figure out how to live and run my part of a lab frugally but effectively.

This book is broken up into two parts: advice for new researchers and frugal use of lab funds. The first section contains a lot of the wisdom that I wish I had when I started graduate school. It includes picking a school, identifying what you're pursuing, and tips for succeeding in graduate school and science. If this describes you, then this portion also contains some food for thought and questions that you should ask yourself as you start your journey into the research world.

---

[2] There were lots of questions about our careers and relationship and how sure we were that a spouse wouldn't interfere with our research and studies.

[3] Which made me really popular with people who wanted kids and a huge concern for professors who thought I would surely never finish and get a degree once I wouldn't stay in the lab until 2am on a regular basis.

I wanted to write the second section to help new researchers think about what they were doing in the lab and why, because if you save money in one experiment, that saved money can go towards another experiment. I think this is one way to improve science and reproducibility as a whole.[4]

I wrote this book with the perspective of a life scientist (because that's what I was), but the wisdom and advice are applicable to anyone running experiments in a STEM discipline (though I may not have much to offer the theoretical mathematicians).

I care a lot about many scientists- I was one (a story for another time). I enjoy the quirkiness, the thrill of discovering something no one else has seen, discussing minutiae in such depth that others think you're crazy (which is a separate problem, I would always tell them), and the satisfaction of accomplishing a difficult task that few others could accomplish (when considering the whole population of Earth). I want you to go off and have fun, boldly experiment, question your assumptions, and not be worried about money; you've got enough to keep you up at night already.

So cheers and good luck!

P.S. For those keeping score at home, I am not an attorney nor a financial advisor. I am someone who has been through this and done this stuff, and this book is about trying to share my experience in the hopes that you will have better research and a better time in science due to the advice.[5]

---

[4] Or get lots of terrible experiments into one paper, but I have more faith in you than that.
[5] See the disclaimer page for additional verbiage.

Who is this book for?

The point of a book is to either help people or entertain people.

This book is for early-stage graduate students, postdoctoral fellows, and faculty to learn how to survive your graduate training and become more frugal in the laboratory.

If you're a scientist, you are a type of entrepreneur. If you're faculty, you have some very similar responsibilities, though grad students and postdocs should notice some similarities in what they do as well.

The responsibility is on your shoulders, like it is for the entrepreneur; you just have additional resources. Accept that responsibility and move forward; the responsibility is yours whether you accept it or not.

This book is also for early-stage graduate students and postdocs who are looking for advice about their choice of school, funding, overall career, and how to be a scientist. I address impostor syndrome and one way to think about your time as a scientist. It is fine if you disagree with my advice on being a scientist, but it is very important that you think hard about the things that I bring up-they will affect your career and your life, so you'd better be sure about your answers.

Scientist	Entrepreneur
Must worry about where funding will come from (grants) and how to use that money efficiently	Must worry about where funding will come from (grants, sales, invoicing) and how to use that money efficiently (which R&D[6] project will I support?)
Must make sure to have a good product (well-planned, well-executed research) that people will read	Must make sure to have a good product or service that people will buy
Must make sure that the product (papers, patents, grant applications) is shipped out on time (to avoid getting scooped)	Must make sure that the product is shipped out on time (to beat the competition or satisfy your customers)
Need to find good staff (postdocs, grad students, undergrad researchers)	Need to find good staff (researchers, technicians, salespeople, etc.)
Increasingly responsible for different legal documents or parts of such documents (grants, non-disclosure agreements, patents)	Responsible for different legal documents (grants, non-disclosure agreements, patents, business registration, trademarks, etc.)
Responsible for marketing (conferences, presentations, selling yourself to tenure committees or professors)	Responsible for marketing (going to relevant conferences, presenting to clients, social media, etc.)

[6] R&D = Research and Development.

Scientist	Entrepreneur
Asking for money (grant committees)	Asking for money (grant committees, bonds, venture capital)
If you fail in your role, the institute shouldn't go away, but your lab does	If you fail in your role, the company goes away
If you run out of money, you're done	If you run out of money, you're done

PART 1: Savvy Advice for New Scientists

This is really for the very new folks, but seasoned scientists might like some of these bits of advice too.

Chapter 1

Graduate School Advice

What am I doing, and why am I doing it? Will this education benefit me?

Why am I doing this?

You need to know why you are applying to graduate school and working to become a scientist. If that's your dream, great! If you want to teach science and know that a PhD is necessary for that goal, then pursue that degree with a clear understanding of the necessary work before you can achieve your goal (and don't despise that work; if you hate doing research and are putting it off, then every day you put research off is another day you put off your dreams and your graduation). If you just want to teach 1 class per week and be on sabbatical the rest of the year while earning (the term is used loosely here) gobs of money, then you should spend time talking to real professors and researchers to understand what their jobs are *actually* like.

If you are just looking for something to do before reapplying to a professional school (medical, dental, physician assistant, etc.), then consider a Master's degree or some other program (or even a job); many professors have been burned by students that the professor put their time, money, and energy into only to lose all of that when the student got accepted to their school and left a pile of unfinished projects lying around. If your transcript reads "pre-med," then you will need to convince the prospective professor that you'll stick around long enough to get the degree (and if you do convince them, actually stick around and finish the job).

If you can't stand failure, then you will either grow immensely in graduate school or run away screaming. Experiments fail all the time for lame reasons[7] or good ones.[8] Failing for 6 months in a row was humbling and not fun to go through, but it built my tenacity and self-confidence when things started to work. It also made it easier to be gentle when teaching others.

If you pursue a PhD, understand that it opens some doors and closes others. You will be considered overqualified for many jobs that you could easily get while working through your undergraduate education. You may graduate and be desperate for a job, any job, and most employers will look at your science PhD and think that there's something terribly wrong with you if you're applying to work there (and you don't get the chance to explain that you're likely fine, but the job market isn't). You are very unlikely to hear back from any job posting that only asks for a bachelor's degree when you walk in with 3 letters after your name instead of 2.[9] Your possible job pool will become more specialized and shrink.

Finally, you can pursue further education after getting a PhD, but you should think long and hard about what doors that additional education will open in exchange for the extra debt and extra years where you don't have a job and therefore don't have income. An

---

[7] Someone contaminated the stock solutions with heavy metals, for example...
[8] For example, the exact opposite of your hypothesis turns out to be the truth.
[9] B.S. or B.A. vs. PhD

MBA with a PhD is only useful if you have a specific plan for it. If you want to go to law school for patent law, look up the starting salaries for most lawyers (it's a bimodal distribution) and figure out how likely you are to get on the best side of that distribution.

In short: think about why you're going to grad school, what doors it will open up, and which ones it will close.

Importance of location and school offerings

Your choice of school and mentor have many effects on your research career. Going to an Ivy-level school has a lot of intangible benefits: you are considered the cream of the crop, those schools do not struggle for money for research nearly as much (grant money, licensing income, and other things go disproportionately to prestigious schools), your alumni network has considerable clout, and prestigious firms (in consulting, banking, venture capital, and more) recruit from your school specifically.

However, the choice of school also dictates living expenses while you go to school, what kinds of support are offered, and what kind of specialty you will have (or be considered to have).

Costs of living

There are places in California that I couldn't afford to live in with a stipend of only $18,000 per year, and a job offer in New York City for $25,000 per year was also going to be very difficult to work with (without loans, that is).[10]

---

[10] These numbers are from real job postings; I didn't make these up.

How expensive is the place you're looking at? Are there less expensive suburbs or spots outside of town that you could live in? Are you planning on going to school in Hawaii, where most things need to be imported in (and therefore cost more)? Is the place famous for its restaurants and, combined with a person's inability to say no to peer pressure, cost more for that person to live in than a small, nondescript college town?

If you're moving to an expensive place solely for your education, then the prestige and likelihood of getting a well-paid job when you're done should be able to pay back those costs; most big-name banks, consulting firms, law firms, and big corporations all recruit from the top schools, so if you want a job that starts out earning over $100,000 per year, you're likelier to find such a job if you graduate from a top school.[11]

However, if you're going to borrow $70,000 at 5% interest[12] to study at a place where your job upon graduation will earn you $30,000 and not grow very much, when will you be able to pay off that loan?[13]

---

[11] Though you are also likely to have significant debt if you go to a top school...

[12] At the time of writing, I see education loans offered with interest rates ranging from 4% to 7% interest. For reference, 5% interest on $70,000 is $3500 per year or roughly $291.67 per month in interest alone. Also, these loans grow while you aren't paying them off.

[13] For reference, see most blogs from humanities majors, where they frequently discuss the lack of desirable jobs (or any at all) in their chosen fields.

Types of support offered

Your school of choice will determine what kinds of support you can get. How likely is it that your mentor will be funded (and therefore you can get a research assistantship)? Will you need to teach? Do you have the opportunity to teach in exchange for income, or are you out of luck? If you're in the sciences, there should be a way for you to get paid to go to school, not to pay to go to school, so make sure that there is one before you sign any dotted lines.

Specialty

Your specialty will be influenced by the professors at your department and your school. For example, one school's program in chemistry may have been influenced (in terms of hiring, firing, grant funding, history, and other things) by the discovery of an important antiviral drug years ago, and so medicinal chemistry may be a big thing at that school and in that department. If that's the case, then there would be an expectation (both within and outside your department) that you, upon graduation, will have more than a passing familiarity with medicinal chemistry. Moreover, your classes may be structured in such a way that medicinal chemistry is included in most of your training and so, even if you really wanted to specialize in materials chemistry, you will spend a good deal of your time studying another branch of chemistry that may not meet your goals.

To further stretch the example, the materials and equipment available in that department may reflect that medicinal chemistry bias and so you might have a much harder time overcoming that bias to study your chosen branch of chemistry.

If all else is equal, pick a place that specializes in the branch of science that you want to pursue. If all else is not equal, put the specialty in your list of considerations.

How are you getting funded?

This section is best for graduate students and postdocs; those of you with full-time jobs should know exactly where your money is coming from and many of these oddities don't apply to you.

Graduate students and postdocs typically earn their salaries in only a few ways:

- Institutional grants, grants to your PI's lab (also called a research assistantship), and/or your very own grant (congrats!)
- Teaching (also called a teaching assistant or instructor)
- Paying your own way (or relatively rich relatives)

Let's discuss the considerations of each path.

Grants and research assistantships

The thing about grants is understanding what your responsibilities are under such a grant. Did you have to sign anything? Do you know what you are prevented from doing under the grant? Who owns the data you generate, any inventions you come up with, or your time? Do you get any benefits personally, or is this all funding for the lab? Importantly, does this sign away your ability to get other grants or other work?

For example, I once was on a grant that prevented me from having any other work- it was a breach of contract for me to teach a class on the side without the approval of around 10 people, many of whom were not at the university I was studying at (or my state, really). The grant came with many benefits (health insurance, some research funding), but I couldn't do any extra work to make ends meet. We made do, but it would be good for you to know beforehand if there are specific stipulations that can affect you.

Is the money taxable? Grant money typically isn't taxable, but you still need to check the terms of the grant and with any grant administrators that manage the grant in question. For example, I had two small grants that, to make it easier on the funders' accounting, simply sent me a check for the grant amount, and then I needed to deposit that check with the lab itself in order to buy supplies with that money. Those grants were considered taxable income (so I couldn't apply the full amount to the lab). However, my stipend from a training grant that I earned was not considered taxable income; therefore, I didn't have to pay taxes on it, but I also couldn't use it for any benefits that require taxable income (e.g., retirement plans).

It can be beneficial to file your taxes, even if well-meaning folks tell you to avoid doing so. If you're supporting yourself and have some minimal taxable income, there are income tax credits that you can apply for and still get some money from the federal government at the end of the tax year.[14] Filing taxes also helps avoid an audit, which is a rare event that, if you win that particular lottery, will be a massive headache.

If you get a certain amount of money toward research funding, make sure that you understand whether it comes from the same

---

[14] Which can either go toward paying bills or invested for later.

pool that your benefits are taken out of- if the "research funding" portion of your grant first pays for your benefits, you could have no money left over for research. Make sure that you know where the money goes and when.

Finally, make sure to look up grants in your specialty and in your department at least once a year (until you're on a grant that supports you). If you are studying heart disease, the American Heart Association does provide training grants (that you must compete for). There are also smaller grants that you can use, which provide two benefits (at least) if you get them:

1. You get the money for your research or living expenses.
2. It makes your resume/CV look better because it indicates that you can think of good research, you can write decently, and that you were looking for ways to be independent.

Teaching

I taught a fair bit as a graduate student- over 3 years, not including the years spent tutoring. In the sciences, you generally get a stipend to live off of and reduced tuition costs in exchange for teaching a class. If you have to pay for the tuition and fees, then you'll need to save up that money each month to cover tuition and fees when they come due. If you're adjuncting, then you're getting a set amount for teaching the class and things are a lot harder to manage.

The benefits of teaching are that teaching can be a lot of fun and help cement the knowledge in your head. I have many fond memories of my students and their antics, questions, and (rarely) obliviousness. I have found that teaching was, for me, the best way to learn a subject, and there are things that I taught to students multiple times that I can't forget (which is a good thing). Helping someone achieve real understanding is also a wonderful experience. If you plan on pursuing a career that involves teaching, then this experience will also help you down the road when you are looking for work.

The drawbacks of teaching are that teaching can take up a great deal of your time and, if you're adjuncting, it's hard to make a living.[15]

Many professors don't like teaching.[16] Part of that is likely due to the time that it takes:

- 1-2 hours to teach the actual class (or 2-5 if you're teaching a laboratory course),
- 10-25 hours per week preparing lectures and slides (at least the first time, but updating lectures takes time too),
- 10 hours for grading (unless you hire people to do the grading, but you still have to double-check that their work is acceptable),
- and another 5-10 hours to be in your office and have no one visit unless you've just handed back a difficult exam.[17,18]

---

[15] This is what is known in the business as an *understatement*.

[16] Which is why you should be kind to those who do love teaching.

[17] I once had a student argue with me for 45 minutes about something she didn't understand but thought her answer was vague enough that it could conceivably mean the right answer.

[18] I also once spent over an hour in my boss' office because a helicopter parent wanted to chastise me for writing a quiz that their child did poorly on.

You, oh young teacher, will also be expected to have office hours, teach the class, prepare ahead of time, grade, and be very patient.[19]

Adjuncting (aka adjunct teaching or adjunct professorships)

Adjuncting usually means that you're teaching a class in exchange for cash, but you are not a full-time employee (so no benefits), you don't get lab space or office space (may vary by school), the competition is fierce, and you don't get enough to live off of for teaching one class; often, adjuncts must teach 3-5 classes per semester to make a decent salary, and no, you don't get reimbursed for wear and tear on your car, travel, or the fact that you usually can't find 3-5 classes to teach at the same school. If you want to be an adjunct full-time, know that it's hard; most (if not all) of the "PhD-on-food-stamps" news stories I've read involve adjuncts.[20]

If you're teaching, the income is taxable. You need to file taxes. No ifs, ands, or buts.

Quick teaching tips and tricks

1. Your syllabus is a kind of contract with the students. Take it seriously.
2. Your syllabus should contain your process for dealing with cheating and you should discuss it on the first day. I have found that dealing with the idea up front clears up any

---

[19] My first lab instructor put it this way: "There are the A students who will earn an A no matter what you ask them to do. There are the B students, some of whom are really sure that they're A students and will argue with you for hours about one point on an exam. The C and D students are really struggling and the F students just do...not...care."

[20] Including the recent news story of a woman who was (is) a prostitute at night so that she could 'live her dream' and be an adjunct. *That's* living the dream? Some things are just not worth your dignity.

misunderstandings and cheating frequency goes way down if people know what is expected of them. You do have to follow through on what you say you will do, however; if you let someone cheat and don't enforce the rules, cheating will become more frequent in your classes.

3. Quickly identify which of your supervisors (or fellow teachers) can help you with awkward situations, e.g., you catch someone cheating, people aren't listening to your instructions regarding open flames,[21] or you need to run to the bathroom during your 4-hour teaching period. You want to know who that person is long before the awkwardness actually happens.

4. Your first time teaching the class takes the most time- you are grading and coming up with exams, yes, but you're also learning the material in a way that you didn't have to learn it before. As such, you'll learn a great deal about the material from teaching that class and that just takes time.

5. Quickly identify experienced teachers. They can help you understand (or cope with) all the strange things that come up.

6. Find ways to enjoy your time with your students (nothing indecent, mind you). I enjoyed the questions that college freshmen would ask as well as the youthful world-weariness of the seniors.

7. If possible, don't take yourself too seriously.[22]

---

[21] "Yes, I'm not exaggerating, please put up your hair so that it doesn't catch on fire."

[22] I once taught a class that required the students to sort through over 70 seashells for one experiment. That experiment happened, one year, to fall on International Talk Like A Pirate Day, so I delivered the entire lecture while impersonating a pirate.

Paying your own way

Paying your own way is a very rare way to go through graduate school in the sciences, but is a very common way to go through graduate school outside of the sciences. You should apply for grants as quickly as possible or find assistantships that come with tuition waivers or tuition assistance. Otherwise, you will need to find a job that can help pay your bills, and that is difficult to balance with the expectations of graduate school.

For most scientists, this doesn't exist as an option unless there has been some sort of misbehavior that was egregious enough to cut the person off from grants or teaching, but not egregious enough to cut the person from the school entirely. These sorts of things are rare and I haven't personally come across any examples (and I hope not to, frankly).

That said, you may find, as I did, that you are still responsible for some portion of tuition and fees. If that is the case, you will have to save up the money throughout the year to pay those bills when they are due- I recommend dividing the total amount by 12 and putting that money aside into a small savings account as a sinking fund each month (seriously).

The FAFSA

Filling out the Free Application for Federal Student Aid (FAFSA) is necessary if you want to apply for federal student loans in the US. However, I'd like to briefly point out one benefit of filling out the FAFSA:

There are small grants that you are eligible for if you fill out the FAFSA. You don't have to take the loans and you can still get the grants.

My teaching assistantship included a tuition waiver, which did not pay for the associated education fees. I used to get small grants that paid part of my fees that I only discovered by filling out the FAFSA, and I avoided taking any of the loans. I saved an additional $1000 a year, and when I was living off less than $18,000 per year, that was a nice amount (over 5% of my yearly income, actually).

In summary, know how the money will arrive for you to go to school and live on (and even use for running experiments). Know what responsibilities you have in exchange for taking that money and be on the lookout for grants, particularly because getting a grant has several benefits for your future.

Chapter 2

Career planning: most PhDs don't become professors

"If anything is certain, it is that change is certain. The world we are planning for today will not exist in this form tomorrow." -
Philip Crosby

Most PhDs are trained by professors to become professors and yet most PhDs will not become professors. No, really. Estimates differ, but these studies typically agree that less than 15% of grad students and postdocs get tenure-track positions,[23] and this number decreases when you factor in how many people drop out of grad school; this also varies by specialty. The situation appears to be happening worldwide, as the numbers are quite bleak for UK, European, and Japanese scientists.[24] Further, this doesn't factor in that if your degree is not from an elite school (top 25%), then your chances of becoming a professor decrease.[25,26] The number of

---

[23] Only 14% of biomedical science PhDs were on the tenure track at 5-6 years after graduation (NIH Biomedical Research Workforce Report, 2015). Note that this number doesn't compare those who did and did not get tenure, so the number who stay as professors might be even lower.
[24] As one example, The Royal Society's 2010 report, "The Scientific Century," indicated that the chance of a PhD becoming a professor was 0.45%. Let that sink in- less than half of 1% of PhDs became professors in their dataset. And we are awarding more PhDs now than we were in 2010.
[25] In computer science, business, and history, the top 25% of schools produce 71-86% of all tenure-track faculty (Joel Warner and Aaron Clauset, "The Academy's Dirty Secret." Slate, Feb. 23, 2015). Look at the faculty where you're at; how many of them have Ivy-league names on their CVs?

available professorships is remaining stable or slowly increasing or decreasing, while we are producing more Ph.Ds than ever. While non-academic jobs (or at least non-professor jobs) are called "alternative careers," the reality is that more than 90% of us aren't going to become professors, and so an academic professorship is the real "alternative career."

Yes, many of you will have professors who don't care to spend time on you if you're not going to become a professor, and this neglect can be hard to take; however, your mentor won't feed you if you can't find a job, so make sure to have a backup plan or three.

So what options do you have? Your options are typically presented as industry, government, or academia. ...right, that's not clear at all. What do these actually mean?

Possible job titles for PhDs

I like lists, so I'll break down these monoliths into lists. These are jobs or fields where I've found postings that listed PhDs as either requirements or requirements to get to higher pay scales, i.e., having a PhD was listed as either necessary or a plus, but it was explicitly mentioned in the posting. This is not a complete or exhaustive list.

Academia
- Professor (tenure track)
- Lecturer/Adjunct Professor
- Research Professor (not tenure track)
- Research Administrator

---

[26] Clauset et al., "Systematic inequality and hierarchy in faculty hiring networks." Science Advances 2015, 1(1).

- Licensing Specialist
- Business Development Specialist
- Laboratory Manager
- Laboratory Safety Officer
- Staff Scientist
- Statistician
- Postdoctoral Fellow
- ...and likely many others

Government
- Regulatory Specialist
- Patent Examiner
- Military Scientist (biology, chemistry, physics, others)
- Agent (FBI, CIA, NSA)
- Smuggling Interdiction Officer
- Government Scientist
- Laboratory Management
- Park Ranger
- Natural Resource Manager
- Fire Management Officer
- Plant Protection Officer
- Nuclear Systems Engineer
- Environmental Engineer
- Systems Capability Leader
- Toxicologist
- Forensic Specialist
- Statistician
- Computer Scientist
- Programmer Analyst
- Hydraulic Engineer
- Water Resources Analyst/Modeler

- ...and likely many others

Industry
- Patent Agent
- Technical Specialist
- Patent Attorney
- Licensing Specialist
- Regulatory Specialist
- Safety Officer
- Laboratory Manager
- Product Manager
- Programmer
- Data Scientist
- Project Manager
- Consultant
- Business Development
- Marketing
- Lecturer/Professor at a for-profit university or educational website
- Trainer for a specific technique or software
- Sales
  - Medical Science Liaison
  - Pharma Rep
  - Biotech Rep
  - Sales Rep
- Customer Service
  - Field Applications Scientist (hybrid of sales, teaching, and customer service)
- Staff Scientist
  - Contract Research Organization
  - Biotech

- o Pharma
- o Product Development
- Financial Analyst (aka Quants)
- Venture Capital
  - o Analyst
- Science Writing
  - o Editor, Journal
  - o Editor, Freelance
  - o Science Journalist
  - o Science Writer
  - o Medical Writer
  - o Textbook Writer
- ...and definitely many others

Here's the thing you should get out of these (incomplete) lists: you have more options than you thought you did. It is not the binary choice of either being a professor or being a hobo under a bridge, clutching your degree to protect it from the rain and sludge, wishing that you'd only listened to your advisor and stayed as a poorly-paid postdoc for another 5 years...

Assess yourself and what you like to do

The first part of getting a job like any of those on the list is to determine what you would actually like to do (or which types of things irritate you the least). You can do this through personality or aptitude assessments, asking your friends and family, or some time spent thinking about your past and what you loved, what you liked, what you tolerated, or what you absolutely would not do again even if they paid you.

I've taken several personality assessments (I once had a job where all the applicants had to take the Myers-Briggs personality test as part of our orientation, for example). Taking the assessments was helpful, but once you notice some themes, it's time to move on and move toward trying out the jobs you think you'd enjoy.

Find tests that assess your skills and strengths. The goal is to figure out what your strengths are, which are much more widely applicable than your research papers for getting into a new line of work and finding it satisfying.

Okay, so you took some sort of assessment (or went with your gut, which is fine too). What should you do with that knowledge?

Play to your strengths

You do need to find something that you would be good at and enjoy, or at least know yourself well enough to figure out what you do enjoy, which may be only a part of the job.[27] What do you derive satisfaction from, and what are you known for on a personal level? Here's a story to help illustrate this:

Before grad school, I was once a janitor at a good auto shop with a good boss. I love to constantly learn new things, which wasn't happening as a janitor; however, I also enjoy the satisfaction of completing a task and completing it well. I was able to enjoy the work a great deal because I focused on that satisfaction, which was present throughout the job. Having my boss tell me that I could be gone for a month before it would start to be really dirty again was a

---

[27] Okay, honestly, it will be a small part of the job. Anyone who tells you about having a perfect job that they love all the time is usually trying to sell you something ("Buy my kit for only $400!").

huge compliment. He knew and I knew that I wasn't going to stay there forever, but I could still be grateful to have a job and have a good attitude because I was playing to my personality (satisfaction from a job well done) and my strengths (I am meticulous with many things, including my research, cleaning windows, and mopping floors; everyone loves a meticulous janitor).

Make connections and contacts

You don't necessarily have to go to official "networking events," where everyone stands in new, itchy "business casual" clothes (whatever that means this year) and tries to hand out as many business cards as possible to avoid ever really talking to someone. Go to things that interest you and decide how many new people you are going to meet- but keep the number low and focus on having real conversations. I remember the people who talked to me for 20 minutes. I can't remember the people who quickly came in, said something snarky,[28] and gave me their card (less than 3 minutes of total interaction); I only remembered that I'd met someone like that.[29]

Quality matters. The people who remember me best are those I followed up with, even when it didn't look like we had anything to offer one another. The future holds many surprises, and that person may be able to help you (or vice versa) in the future.

---

[28] If you only want people to remember you, acting like a jackass is a fine strategy. If you want people to remember you *and* want to do you a favor, being polite goes much further.
[29] I forgot who they were by the time I got to my hotel room and threw the card away because I couldn't remember.

If you want to get someone's undivided attention for a little bit and ask them questions, buy people coffee and ask them fun questions. I like asking the questions that make people smile and light up like Christmas trees as they quickly explain to you what their favorite thing is (when they run out of breath because they're so excited, you've hit the jackpot). You should ask the questions that you really like, but keep in mind that some people want to get right down to business and others don't.

Quality time over quantity in networking- if they don't remember you, they can't help you.

Find resources to help you

If you're at a university, they should have a career center and an alumni network. Find ways to meet with alumni who are doing what you want to do (or at least something related). The career center may or may not be helpful, but they likely have something that can give you a lead on another source of information. Are there local nonprofits that help with skills or job placement? Surely graduated from your graduate program before you did- what are they doing?

Additionally, you have a whole internet at your fingertips. You can find out about the jobs you are interested in if you search for something similar, and then follow up with searches for the exact keywords that you want. It's not as good as a networking event, but you can find out about local events, job fairs, recruitment rounds, and the like.

Social media can work to meet new people. LinkedIn, for example, can be useful, but you don't want to send invitations to everyone who is remotely related to something you're interested in (that's also a good way to get kicked off the site). If you have something in common, you can reference it in your invitation, but make sure to write the person a custom invitation; don't use the default. In any case, meeting new people on LinkedIn is actually harder than going to events and shaking hands- you should prefer to go out and shake hands.

You can use recruiters. Their job is to match qualified people with jobs, which can be of immense help to you. However, you should keep in mind that you are not their client; the company that would hire you is their client. They also make some portion of the new hire's salary in exchange for their services. That doesn't diminish them as a resource, but it does mean that they aren't necessarily going to spend their time helping you- they are spending their time vetting you.

People want to help you but may not know how: help them learn how to help you

Share the things you find with people. Share them with friends and colleagues. One of them may know someone or something that can help you, but they didn't know you were interested in that specific topic before you mentioned it. Your PI may know someone who went to work at a specific company, if you know that company has positions you are interested in. Saying "Biotech" as the place you want to work doesn't work nearly as well as "Company Y's regulatory division."

Heck, tell your loved ones: I told my mother what I was interested in and, in her spare time, she would send me emails with different jobs that I was qualified for. If this is a numbers game, having more people sending you more things that you can do increases your chances of finding a job and something to do with yourself.

"The Side Hustle"

Side hustles have become increasingly popular among the young and the old these days. They used to be called second jobs, but a 'side hustle' generally involves some kind of gig work, i.e., you don't have set hours or a regular place that you should up to in order to work. This isn't the same thing as volunteer work; volunteer opportunities are unpaid but can also have a huge beneficial impact on your career and how you move forward. If you have the opportunity to volunteer for something that moves your career forward, don't be afraid to say yes.[30]

Having a side hustle may be something that your grant funding forbids; if so, stay away from side hustles since you signed a contract that said you wouldn't do any outside work.

For those of you who can do these, there are several benefits to side hustles, including the money (of course) and different kinds of experience. The ideal side hustle is actually one that will move your career forward in some way. For example, editing and proofreading scientific manuscripts will enhance your reputation (if you're a good editor/writer) and be a net positive on your resume for any future applications for scientific work, as well as open doors in scientific publishing. Let's look at a couple of different side hustles and their benefits.

---

[30] I have volunteered as a tutor, camp counselor, treasurer, consultant, nonprofit president, teacher, and counselor. You would be surprised by how often this experience counts in my professional life.

These side hustles will provide money and develop your scientific career as you improve in different relevant skills:

- Freelance scientific/medical writing, editing, and proofreading: can help develop your writing skills and advertise your expertise; direct experience for medical and scientific writing jobs
- Data analyst work: can help develop your data analysis skills and expose you to different workplaces that you might try in the future (e.g., freelance investment data analyst for those looking to get into finance)
- Internships: any available internships in your field can be a huge leg up in getting work outside of academia; clearly indicates that you've been exposed to an industry workplace or the type of work you are pursuing
- Tutoring and teaching: can help develop your teaching skills, public speaking skills, and understanding of the material that you're teaching
- Sales/business development gigs: if you are looking to move into these fields after getting your degree, then this experience can be helpful; if not, this doesn't benefit your career

Side hustles that don't benefit your career (unless you change careers from science)

These side hustles mostly provide you with extra money (nothing wrong with that!), but aren't developing skills that are relevant for you as a scientist:

- AirBnB (or some kind of house rentals/hosting): this could be helpful if you're a social scientist...[31]
- Car wrap advertisements: you don't really talk to many people, so you're not developing any new skills
- Loaning out your car: no new skills are being developed
- Providing rides/ridesharing (Uber and Lyft): develops customer service skills, but those aren't a big part of lab work
- Dog walking and pet-sitting: the skills you develop here don't translate immediately into improved lab work
- Voiceover work: the skills you develop here aren't used in the lab, though this could help you give better presentations
- Online surveys: this could help you if survey design is part of your research; otherwise, not developing any new skills
- Designing t-shirts, coffee mugs, other products: design skills aren't used in a lab except when making figures, and that's a small portion of your time. Your figures will probably be awesome, though
- Developing a dropshipping business: this is developing a side business that takes away time from the lab work; this doesn't develop any new lab-relevant skills, either
- ...and many more

A side hustle can be a huge benefit in terms of making ends meet, reaching some financial goal, and possibly giving you a leg up on the competition during your job search. The main drawback is time- you have to balance the time you give to the side hustle with

[31] "I have put the coffee pot, 3 photos of Jack Russell Terriers, and a copy of a Chinese takeout menu in the bathroom, directly above the toilet. Let's see how Renter 3 reacts and I'll interview them about it later."

the time that you give to the lab, your classes, whatever else brings in the money, and any other volunteer opportunities that come your way.

Get off your behind to get a job

You really have to get up to go find work. You can't just apply online to jobs and never follow up, meet new people, or develop new skills. Sending out applications only through online job boards can lead to lots of disappointment and significant bouts of unemployment. All of my successes (don't worry, my CV isn't particularly impressive) came from jumping at opportunities and not just sitting around. I got collaborations because I knocked on doors. I shadowed physicians because I came to them. I got a research internship because I kept asking people. I became a leader because I did not stop learning new things to try to become better at my job(s).

One of the best pieces of advice I was ever given was this:

Live your life on purpose.

If you passively wait for life to happen, for things to just click into place, they may only do so after a very long period of time, if they ever do. Do you want to be a staff scientist at a big-name pharma company? Okay, find out what people need to do and be like to succeed in those places. What steps can you take today toward becoming that sort of person? What can you learn today to get there? Maybe the place you want to go requires you to work with databases. Can you pick up a book today on SQL? Maybe take a beginner's programming course online?

Maybe the job isn't what you're passionate about. That's fine- just know or learn what you do care about and move forward. Live your life on purpose instead of waiting for that perfect thing to come your way. If you just want to have a job that pays the bills and lets you go motorcycle riding on the weekends, that's cool. If you believe that you just want to pay the bills and go help a local animal shelter, food bank, or soup kitchen, then decide when you're going to do that and don't be ashamed.

Chapter 3

Life happens: marriage, children, and other parts of life

What is failure, and who defines it?
What is success, and who defines it?

Life tends to break our plans, or at least interrupt them. You may fall in love and get married. You may have children. You may lose a job you loved. You may lose a parent or a best friend to a terrible disease in a few years.[32] You may decide that you need to get a dog (or a cat). You may spend a few more years in science and realize that you liked knowing how things worked a lot more than pushing the boundaries of human knowledge and all the meticulous bottle-washing and button-sorting work that goes with it.

Part of personal finance is having a plan that works in good times and bad times; however, part of having a plan also includes being willing to change that plan when life happens.

Let's go for some examples:
- You have a child and find out that, surprise surprise, you really like them and want to spend more time with them. How does that impact your career goals? Be honest, even if the answer is "I don't know."
- You are enjoying your favorite television show when you hear a loud crash. Someone has hit your parked car and

---

[32] I did all of these things in a 7-year span.

driven off (well, as best they can). Will you start budgeting for car replacement or for public transportation? Did you have insurance that would cover this sort of thing?

- You got married (yay for you!). What are your spouse's goals, and do those goals fit with yours? Did your goals change ("I want to just rent apartments forever" becomes "Well, I guess finding a house wouldn't be so bad.")? Have you actually talked about these goals?
- You have trained for years to be in a specific field of study. You get there, get your PhD, and find out your skills are now considered worth less than when you started and your position is cut (i.e., you're laid off). Do you have any savings? What do you do now?
- You get your PhD and develop a life-threatening allergy to something you used to work with all the time, so you can't do the same projects you were trained to do. Additionally, it's becoming clear that no one wants to train you in something new. What do you do now?

Determining your personal goals and visions is wonderful. It can be a tremendous help in identifying what you want to do. It is wonderful to figure out what you want your life to be about.

I got married and had children while in graduate school. For me, it was a great motivator to be efficient at work and get out of school with a degree. I also figured out that being a professor doing high-powered research wasn't as important to me as being a great parent and spouse. According to some, I am a failure because I didn't go become a professor (or I "just proved that [I] couldn't have made it anyway"). My current job is to help scientists instead of being one, and I have more time to be a good dad when I go home.

What is failure, and who defines it?

The point is that you and your circumstances will change. Your desires and dreams may also change. That's okay. If you change, you don't have to follow old dreams just because you had those dreams at one time.

If you decide that something else will make you happier than your first goal, you are not a failure or a traitor to science. How do you define success? If someone followed an old dream that now makes them miserable, do you consider them a success? Or would being successful mean being the person who decided on new dreams and is now happy, even if the new ideas are much less impressive to their old friends?

Who defines success? Typically, you. You will be around yourself in 20 years, and the people who want to decide your life for you probably won't be around in 3 years, much less 5, 10, or 20 years.

What do you want your life to be about? Science can be tremendously fun. Science, the process of slowly uncovering the way something works, is wonderful- I miss doing the meticulous work, I miss talking to people about esoteric tidbits over coffee ("Did you see the latest J Neurosci paper talking about…"), and I miss the constant learning...but my goals and circumstances changed. Don't be ashamed if the science isn't more important than something else, e.g., your kids' health being a higher consideration than the science, wanting to work with animal shelters is higher on your list than collecting tadpoles this weekend, or having children at all being a higher consideration than a high-powered, high-social-status career. Such things are decisions that only you can make (or you and your spouse, if you're married).

We live in a very privileged age, historically, in which we can decide on what kind of work to pursue because we actually like the job, instead of just struggling to survive and taking what you can get in order to eat. I am grateful to have been a scientist.

A scientist who didn't or couldn't stay in the laboratory is not necessarily a failure.

Chapter 4

Scientific advice

Read widely...

"Chance favors the prepared mind." -Louis Pasteur

The Pasteur quote above is one of my mentors' favorite quotes. It still holds true. If you're beginning in research, you have to make time to read. If you're continuing in research, you must make time to read. The day can easily slip by, and so can the week, the month, the year. Learning from your own experiments is great, but inspiration can be hard to come by if you're facing the same problem week after week.

Has someone else faced the same problem? What about a similar one?

Maybe they're not in your field. I optimized one of my techniques based on work in a completely different field and fixed a huge problem by using a solution devised by someone in a separate line of work. If you read widely, you are exposed to ideas that aren't in your immediate vicinity, you learn about the multiple ways that different researchers approach a problem, and you'll have more ideas about how to tackle novel problems. There are real benefits to multidisciplinary work.

When I wanted to learn about a new topic, I would start by printing out around 30 papers (this habit was only reinforced by my preliminary exams) to read and get ideas about a given field. We can do this more easily than any generation before us due to the various search engines that now exist; you don't necessarily have to walk into any library at all. If you don't know how to get at the papers you want, there are librarians at your university who get paid to help you (and are likely very helpful people), so take advantage of the opportunity!

...talk widely...

"Many ideas grow better when transplanted into another mind than the one where they sprang up." -Oliver Wendell Holmes

Ask people why they do what they do (including the steps of different scientific protocols). If they don't know but say they learned from someone else, ask the person that your colleague originally learned from. Dig to the roots of the issue and find out what evidence people are resting on for why they perform specific techniques. Make sure to do this with research papers, too; it is worthwhile to get down to the very first paper in a long chain to see if you agree with the evidence or idea that the authors based their next 5 papers on.

For techniques, remember that some parts are crucial for the technique to work, and some parts are not; science is a human endeavor carried out by human beings, and it's unlikely that we are perfectly efficient and have no superstitions or wasted steps.[33]

---

[33] "Look, I'm sorry, but the microscope doesn't work without playing gangsta rap on the radio. It can't be helped."

Optimizing a technique includes determining what steps are crucial and why.

If you're in biology like I was, go make friends with someone who studies physics. Bell Labs encouraged scientists from different fields to interact and share ideas; go get lunch with people who work on different things. The life sciences world is becoming more interdisciplinary, so go learn how to live in an interdisciplinary world. Your esoteric problem may be someone else's everyday irritation.

...think thoroughly...

"Science is a way of thinking much more than it is a body of knowledge." -Carl Sagan

Take time out each week to just think. This is a luxury afforded to scientists and people in such white-collar occupations (and the point of the book *Deep Work* by Cal Newport). You are being paid for your brain, so take the time to think about what you are doing, why you are doing it, and what can be gained from your time and attention.

- Are there suitable controls in your experiment?
  - Example: You are staining for a particular protein. Your control is another tissue that is known to express that protein. That's a good positive control; do you happen to have a negative control (where you shouldn't see that protein)?
- Are there better controls that you could use? How could you get them?

- Example: Your experiment could use these really awesome knockout mice to show that you're actually looking at the protein in question. Do you know any labs that have that mouse and could give you some mice? What about just sending you frozen tissue?
- What hypothesis are you really testing in your experiment? Are there possible confounding factors?
  - Example: You're treating cells with a particular drug. However, when you add the drug to the cells, you also replace the cell medium with fresh medium. Are you testing the effect of the drug, the effect of fresh medium, or both together? If the drug must be diluted in DMSO, do you know what the effects of DMSO are on your cells?

If the weather was nice, I would take a walk outside and calmly stroll in the sunshine. The undergrad students would hurry past me, but I would just take in the sights (leaves and tree limbs gently swaying in the breeze, sunshine on my face, clouds in the sky, or a newly-planted flower bed) while thinking about a particular issue I was facing in the lab. The best days were those where I'd invite the postdocs in the lab next door or a friend to have some coffee, and we could all walk together to the coffee shop, while discussing different issues we were facing. There were plenty of times where I'd end up giving someone else a paper that pertained to their problem or receiving advice to help me with my experiments.

If your supervisor/mentor/boss says, "I'm not paying you to take a walk!"

...then one (admittedly impolite) response is, "Yes, you are. You are paying me to solve problem X, and I am thinking about how to solve problem X. One of the best ways to spur creativity is to go outside of your normal environment."

If you say, "I have too much to do, I don't have time to think about things!"

Your career could very well depend on solving your specific problems, and having a PhD ideally means that I could plop you at a desk, give you a huge problem, and you should be able to come up with ways of solving that problem (or figuring out what resources you have to solve the problem if you can't do it by yourself). Have you had the same problem for 3 months? Go think about it! How could you get help (colleagues, biotech rep, or the library)? What resources do you have that could help out (specific books, an old friend, a separate industry that faces this problem routinely)? If you don't think about it now, will you do so in another 3 months, when 6 months have been spent on a single problem?

...remember who you are...

Impostor Syndrome

> *"Turning pro is a mindset. If we are struggling with fear, self-sabotage, procrastination, self-doubt, etc., the problem is, we're thinking like amateurs. Amateurs don't show up. Amateurs crap out. Amateurs let adversity defeat them. The pro thinks differently. He shows up, he does his work, he keeps on truckin', no matter what." -Steven Pressfield*

Impostor syndrome is pretty common among graduate students, both men and women (one report indicated that 100% of the surveyed women responded that they felt like impostors). Regarding impostor syndrome, I know that it's hard. You feel like you're the only person who doesn't get it, or that everyone else belongs and you don't. Everyone else is more accomplished, more with it, more together. They have more papers, they have more money, their advisor is more famous, and no one else has ever had these same problems.

That's typically a lie we tell ourselves.

If you feel like an impostor, did you:
1. Lie to get into graduate school?
2. Fake your test scores?
3. Cheat thoroughly in undergrad courses?
4. Get a fake passport and apply under fake names?
5. Wear a form-fitting mask every day so that no one knows your true face (cue creepy music) and hire consultants to give you ideas for experiments?

If yes, then you are an impostor.

If the answer is no, then you're a young researcher, <u>like the rest of us</u>.

If you don't have as many papers as the person next to you, it is likely just fine. I got 4 papers out of my grad school career- a respectable number, but nothing spectacular. I met a visiting speaker who got 8 out of grad school and bragged about it to the room. It was easy to feel deficient in that room with that speaker.

However, I stretched the money I had available to make things happen, was allowed to plan my own work, and wrote thoroughly. I initiated multiple collaborations to get things moving.

The person with 8 papers was in a well-funded lab with a mentor who planned out their experiments for their students.

I remember speaking to someone who was in a lab where everyone in the lab was included on every paper, no matter how much or how little they contributed. It's a rough ride when you're being quizzed about a paper you contributed very little to and your ignorance is being revealed.

The grass may not be greener on the other side; it might be Astroturf™.

Your value is not wrapped up in the lab or the school you went to or your PI's opinion of you. You likely have loved ones who don't care how many papers or grants you have, whether you ask the right questions at talks, or care how famous you are in your small pond of science; they care about how you make them feel.

I believe that you have intrinsic value as a human being.

Remember these things when you feel like an impostor.

Suicide

News stories are produced on a regular basis about the suicide rate among young researchers; various studies peg the numbers around 10% for the number of both male and female graduate students that have considered suicide,[34] around 7.5 per 100,000 commit suicide,[35] and over 40% show depression and other symptoms that may be comorbid with suicidal thoughts.[36] For comparison, 3.7% of the US population has considered suicide, 15 per 100,000 commit suicide every year,[37] and 6.7% have depression.[38] Taken together, it appears that there are greater rates of thinking about suicide and greater rates of depression in graduate students than in the general population.

To relate what happens after someone commits suicide, I have only my own experience to share. I'm still hurting from two very important people in my life committing suicide. It has been two decades and I still pick up the phone with excitement and think, I should tell them about this, they'd be so happy...except there's no one there to call any more, and my excitement is replaced with a quiet sadness.

One of my favorite quotes is this:

"If you want to feel different about a place, take someone you love there." -Louis Armstrong

---

[34] UC Berkeley Graduate Student Happiness and Well-Being Report, 2005.
[35] Silverman et al., "The Big Ten Student Suicide Study: a 10-year study of suicides on midwestern university campuses." Suicide Life Threat Behav 1997, 27(3): 285-303.
[36] UC Berkeley Graduate Student Happiness and Well-Being Report, 2015.
[37] Crosby et al. "Suicidal thoughts and behaviors among adults aged ≥18 years --United States, 2008-2009." CDC, 2011.
[38] National Institute of Mental Health, 2017.
https://www.nimh.nih.gov/health/statistics/major-depression.shtml

Unfortunately, this also works in a negative way- there are places that I loved that now carry the stain of the suicide on them, replacing all those positive emotions with grief when I pass something as simple as a public park.

Suicide ends one person's pain and spreads it further out. Despite what the stories are on television, there is no redemption in your suicide, you aren't remembered as a hero, you don't suddenly receive the love and accolades that would make you happy, there is simply emptiness- that void where you once were- and the grief. Marriages and families have shattered over the deaths of children and suicides.

If you are thinking of killing yourself over grad school, find help immediately or get out of grad school! You can do both, if necessary. It is far better for you to get out of grad school than to stay and kill yourself. If you're thinking of killing yourself over the pursuit of or not getting tenure, get help and go find another job! It is far better for you to leave the academy with outstretched middle fingers than to leave a corpse in the lab.

You are more valuable than you think you are. The academy can burn and you'd still be valuable.

I don't claim to know the depths of anyone else's sadness; I can only say as someone who survived loved ones killing themselves, that there is likely someone who cares enough to stay and listen to you. Go get help, go see friends, go see family, call suicide hotlines, go see therapists, take advantage of university mental health resources, just do what it takes to let someone know and get yourself some help.

...and look for opportunities

You have opportunities to do different things right now. Some opportunities you have to make; I have stories of knocking on doors and asking people for help or their time, and I went a lot further than expected because of those knocks.

Do you want to get involved in business when you're done? See if your PI is okay with you taking classes to get an MBA while pursuing your PhD Not every university allows it, but you don't know if you don't ask. I've seen a university pay for someone to get concurrent degrees, so you could walk out with two advanced degrees and get to say goodbye to school forever, instead of coming back later.

Is there a visiting speaker in another department who has a job that you want to learn about? Plan out how to go see that talk and ask them questions later. I once had an hour and a half to ask a visiting speaker questions about their job solely because I jumped at the opportunity; the other people in the room asked a couple of questions and mostly looked at their phones, while I looked the speaker in the eye and asked about their experiences.

Does your university include a certification program in something you're interested in? Go find out how to get involved. Does it cost you anything extra? Are there grants that you can apply for to get one of these certifications?

There are opportunities that are only available when you're a graduate student, and some that are only available when you're a postdoc, and some that are only available when you're a faculty member. Identify what you can do now and what you can make time to pursue.

Chapter 5

Benefits are important

"You don't know what you've got 'til it's gone." -Joni Mitchell, "Big Yellow Taxi"

Workplace benefits are crucial if you can get them; when I was in graduate school, these were difficult to come by or nonexistent until you had your degree and were no longer a postdoc. When you're a student or postdoc, getting benefits is a huge plus; if you're faculty, they are necessary, i.e., you need to have very good reasons for taking a job without them. Benefits will come out of your paycheck (or your grant funding), but they will vastly increase your quality of life. This will be US-centric; my apologies to my non-US readers.

Health insurance

Health insurance is the big benefit and the most obvious; many graduate students and postdocs lament not being able to just go to see a doctor when they are sick. If you can get on a group plan, you should; the uncertainty in the US health insurance market has made it expensive to get your own plan. Even after subsidies, I have been offered plans that cost more per month than I made per month. Any sort of group benefit through your workplace is likely to be cheaper and more comprehensive than what you can get on your own.

Here are the things that are worth checking out in your health insurance plan (in short, the whole thing):

- Co-pays (the amount that you will pay, as your share, at the doctor's office or the pharmacy for your prescriptions)
- Deductibles (how much will have to come out of your pocket before the insurance company starts to pick up the tab?)
  - Is the deductible high enough to qualify you for a health savings account (HSA)? For single people, the minimum deductible is $1300 per year and is $2600 for families in 2017.
- Out-of-pocket maximums (how much will you be asked to pay in total for your whole family per year?)
- Benefit maximums (how much will the insurance company pay in one year? This is more of an issue with chronic illnesses)
- What procedures are covered? Are you expected to pay to visit a specialist?

If you plan on having children during your time at that particular workplace, check what kind of maternity coverage they have. During my time as a graduate student, I needed to pay for all maternity care out of pocket when the insurance company decided to drop maternity coverage.

Health insurance may come out of your paycheck or out of your grant funding; check how it will be paid in order to understand whether your paycheck or your lab will suffer paying your insurance.

Dental insurance

You should get dental work done, period, during your time in graduate school or postdoctoral fellowship.[39] Dental insurance, however, should be looked at with a skeptical eye. Most dental insurance covers 50% of the costs of dental work, which can often mean that you're spending more on having the coverage than you're getting out of it. If your coverage will cover more, then just be sure that you're not paying too much for it, e.g., if your insurance covers tooth cleaning and that's it, a cleaning costs $75, you get two cleanings a year ($150 total), and you pay $30 a month for it ($360 per year), you are overpaying unless you get some other benefit.

When we had no dental insurance, we still saved the money to get at least two cleanings per year. That office cut us a deal because we paid the whole balance off in one check and the dentist's office didn't have to go through the hassle of dealing with an insurance company.

Vision insurance

Vision insurance can have the same pitfalls as dental insurance if you're dealing with glasses; I keep pairs of glasses for years, so that benefit only comes up once in a while, and the cost of visiting the optometrist might not be that high. If you get contact lenses, though, the cost of your insurance covering it could work out quite well for you.

Most vision insurance does not cover laser surgery to get rid of your glasses. It should be noted, however, that the rate at which

---

[39] A good friend of mine went without dental checkups until their tooth abscessed and they were in tremendous pain. Please don't spend time doubled over in pain because you didn't want to pay to see a dentist. The root canal and associated surgeries cost more than the costs of all the missed cleaning sessions.

your vision gets worse seems to plateau around age 50, so if you get laser surgery at 25, you will likely need it again later. Also, if you have astigmatism, laser surgery won't work on your eyes (at least for now, but the technology is improving).

Life insurance

"Insurance is the only gamble where you lose if you win and you win if you lose." -Mad Magazine

I have a family. I have a young family that would be in a lot of trouble if I suddenly died. As such, I do keep life insurance and got it as soon as my first child was born. Some workplaces offer life insurance as a benefit; I signed up for the benefit, but it was still too meager to take care of my family if I died. As such, I signed up for another plan outside of work.

Life insurance is a benefit that is paid out if and when you die during the terms of the life insurance contract. If I die young (while I'm covered by the insurance), then my surviving family will receive some money to help them pay the bills and survive.

Any life insurance is legalized gambling; when I buy a life insurance policy, I am gambling that I will die within the specified timeframe and will pay much less than the insurance company will pay out, while the insurance company is gambling that I will survive the term and they will take all the money as profit.

The types of life insurance currently available are:
- Term
- Increasing term
- Decreasing term

- Permanent
  - Whole life
  - Universal life
  - Variable
  - Variable universal
- Survivorship
- Final expense (aka burial or funeral insurance)

Term life insurance is good because it's simple to understand and cheap; if the premiums are paid and the insured person dies, the agreed-upon amount of money will be paid to the beneficiaries (the people who get the money; in my case, my wife and children) if it happens within the term (5, 10, 15, or 20 years or so). They are cheap, effective, and do what they say on the tin. If you keep getting these as you get older, they will get more expensive.

Increasing term life insurance means that the payoff amount increases as time goes on, whereas decreasing term life insurance means that the payoff decreases as you get older (but the payments likely won't change).

Permanent insurance doesn't have a term but instead goes on until the insured person dies. Most permanent plans have some sort of savings or investing side, but it's usually a losing proposition for you; if your surviving spouse gets the agreed-upon payoff amount but none of the savings or investments, they are usually losing out. Alternatively, they can just get the savings or investment component, which may be less than what you thought was the payoff amount.

Whole life involves a savings component that can grow over time and eventually, possibly[40] provide dividends to you, but the savings grows at a rate that is set by and benefits the insurance company, not you.

Universal allows you to have more control over where the payment money goes (your benefits or your savings), but remember that you don't set the interest rate on that savings account.

Variable and variable universal are similar to whole life and universal except that you can invest in stocks, bonds, and the like instead of just the savings account. Typically, the insurance company gets the biggest share of any benefit that your investments produce.

Survivorship insurance involves multiple people and can vary by when the payoff actually occurs. It may pay out when person 1 dies or when person 2 dies.

Final expense insurance can be quite useful for someone in a bad situation, like a prepaid funeral. If you're taking care of an ailing parent who doesn't have a pair of nickels to scrape together, prepaying the funeral or having a way to do so can be very helpful-those grieving don't have to find the money to pay for things. As someone who has needed to find a way to cover an unexpected funeral,[41] it's pretty stressful to find the money (in your accounts and any you've been given authority over) to pay for the funeral

---

[40] It needs about 4 more hedging words, but that would be really dull to read. Basically, it's very unlikely that you'll get a real benefit to the savings component.

[41] Though I suppose that an expected funeral is the kind of thing that lands someone in prison...

happening in 2-3 days in addition to trying to grieve, clean things up, get all the legal paperwork done, and comfort others who are grieving.

I prefer and strongly suggest simplicity- in this case, term life insurance. Who benefits when you combine your life insurance with a savings or investment plan? The insurance company, typically. Let your savings be your savings and your life insurance be your life insurance. I have not yet seen one of these Franken-plans that try to be two things at once succeed at doing either thing well.

Insurance type	Description	Recommended?
Term life	Pays specified amount when person dies	Yes
Increasing term	Payoff amount increases with time	No
Decreasing term	Payoff amount decreases with time	No
Whole life	Combines insurance with savings	No
Universal life	Combines insurance with savings	No
Variable	Combines insurance with investments	No
Variable universal	Combines insurance with investments	No
Survivorship	Pays when specified people die	No
Final expenses	Pre-pays funerals and related costs	Maybe

Retirement plans

I do suggest seeing where and what your employer contributes to your retirement plans (if at all). If they contribute to your retirement plan, contribute enough to get the full employer contribution- a 5% employer contribution is like getting a 5% raise immediately. Just remember that you want to invest in stocks, bonds, and things like that- annuities are only really useful when you're already retired.

Check what options you have at your workplace, and if they're really terrible, just contribute enough to get the free raise and put the rest of your investment money elsewhere.

PART 2: A Savvy Scientist's Laboratory

Introduction

"The way to wealth is as plain as the way to market. It depends chiefly on two words, industry and frugality: that is, waste neither time nor money, but make the best use of both. Without industry and frugality nothing will do, and with them everything." -
Benjamin Franklin

"He who will not economize will have to agonize."-Confucius[42]

Why run a lab cheaply?

Why run a lab cheaply? Well, the basic answer is that if you are efficient with your money, you can run more experiments on the same amount of money.

I've found that this can often be overlooked in the name of simplicity, e.g., "Yeah, I use twice as much of this reagent as I need to because the math is simpler." However, let's say that you are using an $80 reagent that, when you use twice as much as you need, only lasts for 2 weeks. That's $160 per month.

If you use only as much as is necessary, then that 2-week supply becomes a 4-week supply. You just saved $80 by taking some additional time to plan out what you were going to do (in my experience, 10 minutes at most of extra planning time, the first time you do it). What could you do with an extra $80 (at the time of writing)?

- Buy 4 new lab coats (can be found for $20 or less per coat)

---

[42] Every time I've seen this quote, it is attributed to Confucius, but the rhyming *in English* is a little too convenient.

- Get over 1000 disposable pipette tips
- Get at least 25 grams of agarose (I used to make gels with 0.25 grams of agarose, so that's 100 gels)
- Buy a treat for a lab (that's at least 4 large pizzas or a lot of coffee)[43]
- Save it for two months toward a cheap power supply for electrophoresis
- Cover additional mouse colony costs that you could put toward increasing the number of samples in your experiment (make your statistician friends a little happier)[44]

Many grants and grant award systems 'reward' you for spending all the money in your grant; in the past, no one was really rewarded for being careful or efficient with the money, and you might have gotten less funding in later grants for being thrifty. I understand that kind of mentality is hard to put away; however, I suggest thinking of it in terms of additional experiments.

Additional experiments allow you to run additional controls, increase sample sizes, or add new experiments with the same amount of money. All of these things improve the science and improve the manuscript that you want to send out. These things make your work stronger and better able to compete for space in higher-tier journals.

When the NIH budget was doubled in 2003 under President G.W. Bush, that allowed many more scientists to get funding and

---

[43] Free food and free coffee are the siren songs of laboratories; you can lure unsuspecting researchers into conversations and friendship this way. Think of it like a really polite mousetrap.

[44] Just kidding. There's almost nothing in biology that makes a statistician happy.

become professors; since then, the funding growth has leveled off, but we are still awarding many new PhDs each year (about 54,000 PhDs in 2014, ~12,500 of which are in the life sciences, and the number has increased since then).[45] With that in mind, there are many more hands trying to compete for a pool of funding that hasn't increased much in size since 2003 (at least enough to offset the additional hands).

If we're not in trouble yet, then my opinion is that the day is coming when saving money is mandatory, and therefore I'd save money anyway because...

"It's tough to make predictions, especially about the future." -Yogi Berra

---

[45] National Science Foundation, "Doctorate Recipients for US Universities: 2014" and "Survey of Earned Doctorates, 2015."

Chapter 6

Planning, by day and by experiment

"A goal without a plan is just a wish." -Antoine de Saint-Exupéry

"In preparing for battle I have always found that plans are useless, but planning is indispensable." -Dwight D. Eisenhower

Plan your experiments and your day- your time is precious

Okay, so you're with me about saving money. "But if I do things the way you describe, I'll be at work until 10pm!" After all, your time is precious and you don't ever get it back once it's gone.

The answer to that is thorough planning of your day.

Thorough planning isn't always a popular topic or idea, since many supervisors (in any industry or setting) equate the time you spend physically in a place ("seat time") with productivity. True productivity, however, is actually getting things done. If you were at work only from 8am-5pm but got 4 experiments done, were you less productive than the person who was at work from 9am-8pm and did the same 4 experiments?

Nope! So plan out your day and what you will do.

I've found that planning out by goal and by time are helpful, and planning helps me remember if I need any common-use equipment for the day; if I do, and it's all taken, then I know that early in the morning and can re-evaluate what I want to do today. For example, let's say I have 2 quantitative PCR experiments that I need to run on a common-use machine. Each run will take 2.5 hours to run and half an hour to set up (thaw samples, load plate, etc.). Therefore…

8-9am: Plan out my day (15 minutes, tops), have coffee (ongoing), answer emails (selected ones, anyway…). Did I sign up to use the qPCR machine? Go sign up if I haven't.

9-9:30: I signed up for this morning, so I need to get started now. Thaw out my samples and calculate how much I need of each reagent.

9:30-9:45: Load samples into a plate, load the plate into the machine.

9:45-10: Program machine (maybe I'm just slow at typing), run machine.

-The samples will be done by 12:30pm, so…

10:15: Additional experiment planning, go take a walk outside to think.

11: Have lunch; invite a friend?

12-12:30pm: Thaw samples, use the calculations I made this morning, load samples into a plate.

12:30: Load plate into machine, program machine as quickly as I can, run machine.

-If I take another 15 minutes to program the machine, then my plate will be done by 3:15.

12:45: Analyze data from the first run.

1:30-2: Meet with a graduate student from another lab. Bounce ideas off each other.

2-3: Read 1-2 relevant papers.

3:15: Remove plate, take data from machine to analyze.

3:30: Go to a talk that I am actually interested in.

4:30: Talk is done, scoot out through the back door.

4:30-5: Plan part of tomorrow, analyze data; maybe I'll start with data analysis tomorrow morning. I may take some of the data analysis home.[46]

It's certainly not the sexiest part of science, but this can help cram more experiments into your day (I certainly moved things around on some mornings so that I started work before 9am) so that you can do some of the things you want to do after 5pm. And if you want to stay later, you can get the next experiment started.

We can compare this list above to…

8-9:30am: Have coffee (ongoing), answer emails (all of them, maybe?), check social media, re-read my lab notebook to remember the next thing to work on.
9:30-10: Make calculations for qPCR plates, remember that I need to go sign up for the machine.
10: Somebody else signed up for this morning. Need to wait until 1pm before I can use it. Go check social media or get lost in a website while I wait...
11am-12pm: Go get lunch and eat.
12:30-1: Thaw samples, load samples onto plate, load plate into reader once previous person's run has completed.
-This run will be done by 3:30pm. Do I skip the talk I wanted to go to and run my next plate, or go to the talk, run out after 4:30, have samples ready to go by 5, and the plate is done by 7:30pm? Should I have just signed up to do this tomorrow and figured out something else to do today?

---

[46] I enjoyed analyzing data in less-professional attire and without shoes. A good beer or cup of tea can also help.

Planning and checking things at the start of the day isn't exciting. But it does prevent headaches later,[47] which are much easier to deal with at 9am or so instead of after you start prepping samples.

If planning out the whole day seems overwhelming, then you can start by simply writing down a checklist: what are the things that I need to accomplish today? It can be helpful to note what things must be done today and which things would be nice to get to, but aren't as necessary.

Penny wise and pound foolish

"People who are frugal understand the value of a dollar and make informed and thoughtful decisions. People who are cheap try to spend as little money as possible." -Frank Sonnenberg

Something to note for all these cost savings is to avoid being penny wise but pound foolish, which means that you're being cheap on small things but wasteful or foolish on big things. I want to teach you about making informed, thoughtful decisions; this book is not called "The Really Cheap Scientist." Part of being a scientist is making informed and thoughtful decisions about your experiments and the results; I am just trying to extend that lesson to finances as well.

Let's go through a couple common iterations of being penny wise and pound foolish.

---

[47] Like discovering that someone else had used almost all of your reaction buffers. If you had prepped your experimental samples beforehand, the samples would have been ruined by the time you made the replacement buffer.

Equipment that won't meet your goals

As an example, let's say that you're thinking of purchasing a
device that you want to move only a few nanometers at a time;
however, the devices that will move only a few nanometers at a
time are quite expensive, and so you get a great deal by buying one
that must be cranked by hand...except most people can't hand-
crank something to gently move it only a few nanometers. In this
example, yes, you saved a lot of money by not getting the
expensive device, **but you can't do the experiment**. That is being
penny wise and pound foolish, i.e., you saved money but
completely defeated the purpose of buying the item in the first
place.

Another example: You bought a microwave at a garage sale that
you want to use to boil agarose for making agarose gels. It turns
out the model isn't powerful enough to boil the agarose in any
convenient time frame ("Oh, it's just a 40-minute wait! No
problem."). You got a cheaper microwave, yes, but not one that
will do what you need the microwave to do. I did once observe
someone using a garage-sale microwave to generate blue sparks
while trying to heat up agarose; we never used that microwave
again and just heated our agarose on a hot plate.

Substitutions that actually change the research question you're
investigating

What on earth does this mean? An example might be effective
here:

You have an experiment where you will run a rat through a maze
while treating one brain area with an ion channel blocker. The
channel blocker you are interested in (SelectiveToxin) costs $140

for 3 mg, and it specifically blocks one ion channel type. You can get another blocker (DirtyToxin) for $50 for 5 mg, but it is less selective: it blocks three ion channels. If you use DirtyToxin in your experiment, you are testing whether any of the three ion channels it blocks are involved in the maze-running experiment (including combinations of the three); if you use SelectiveToxin in your experiment, you are testing if just the one ion channel is involved. For a real-world example of this, gamma-DGG (ɣDGG) is a broad-spectrum ionotropic glutamate receptor antagonist (blocks AMPA, kainate, and NMDA receptors), and NBQX is a selective AMPA receptor antagonist (a type of ionotropic glutamate receptor).

Notice how changing one thing to save some cash left you with different experiments?

As another example: You want to determine the effects of compound A on muscle tissue. Compound A is relatively expensive. Compound B is cheaper and structurally similar, but different enough from compound A that you can't say the effects of one will be the same as the effects of the other. If you run the experiment with compound B, then you're testing what compound B does on muscle tissue, not what compound A does to muscle tissue. For a real-world example of this, testosterone and estradiol are very similar in terms of their structure, but they have very different effects in the body. If you just test the effects of estradiol, can you really claim that testosterone must surely have the same effects because their chemical structures are similar?[48,49]

---

[48] Don't even think about it- the answer is no. There are lots of papers in the medical literature that point out the different effects of estradiol vs. testosterone.

Lesson: don't perform an experiment that doesn't address your question, even if the other idea is cheaper. If you want to know what compound A does on this tissue type, then don't use compound B or a different tissue type because compound A is expensive; if it doesn't answer your research question, you can't get the time or money back. If you want to know about compound A, use compound A. If you want to know about what a cat does in response to someone throwing a stuffed octopus, then use a stuffed octopus and a real cat, not a catnip toy and a plush cat doll (though I'd love to see that paper).

With equipment and consumables, even if it's a deal, make sure that the item can accomplish what you set out to do in the first place. It can help to think of the negative or opposite terms:

You: "I have this thing I want to buy. It'll solve [big ugly problem] for us without any difficulty."

Think about:
1. Who will use this? I trust myself, but will the person using it work with it as easily as I can?
2. What are some easy ways to screw up the job when working with this?
3. What are some easy ways to screw up the device itself?
4. How likely are scenarios 2 and 3?
5. What does this thing actually do?
6. What research questions will this thing actually test?

---

[49] And we're not even getting into stereoisomers or chemical chirality yet.

Can't finish the experiment

If you haven't fully planned out the experiment and run out of money in the midst of it, then it doesn't matter how beautiful, elegant, or powerful the experiment was- an unfinished experiment doesn't get reported, published, or publicized until someone finishes it. Make sure that you can finish the experiment and that you can finish it well, and that the materials won't go bad in the freezer while you wait for someone, somewhere to finish your experiment.

Plan out the total cost in terms of reagents and time and make sure that you can finish what you started. If it's a razor-thin margin of error, either allocate more resources or re-think the experiment, because something always goes wrong (see Murphy's law). I always added a 10% margin of error to my experiments and was pleasantly surprised when I came in under budget (or pleased that I factored in the extra cost when something came up and I could still finish my work).

Chapter 7

Cheap equipment (machines, glassware, etc.)

"The essence of science is independent thinking, hard work, and not equipment. When I got my Nobel Prize, I had spent hardly 200 rupees on my equipment." -C.V. Raman

"Low budgets force you to be more creative. Sometimes, with too much money, time and equipment, you can overthink." -Robert Rodriguez

Equipment is one of the chief concerns when setting up a new lab (or a new functionality in an existing lab, e.g., adding a confocal microscope to a lab that specialized in molecular biology without any microscopes). It can require a significant capital outlay at the beginning and maintenance costs as time goes on. There are multiple ways of saving money on equipment, but they all have pros and cons. The methods that we'll cover here are:

- Get multiple quotes
- Buying demo models that the salespeople have brought on the road to show customers
- Buying refurbished/remanufactured/reconditioned equipment that was likely repaired with original parts
- Buying used equipment on eBay and similar auction sites
- Building the equipment yourself (which has a long and storied history in science)
- Getting the equipment for free or close to it (may involve dumpster diving)
- Doing the maintenance yourself

Most of the cheap equipment that you can get is used. Buying used is not universally attractive to people, but it can be a viable option when all the factors are considered (would you like to save up to 70% on equipment?).[50] Here is a (not exhaustive) list of questions that you can ask yourself about buying a piece of used equipment:

1. How sensitive is this application?

For example, you may be looking at purchasing an electron microscope. You don't have the expertise to build some equivalent DIY contraption, it's something that the lab will work with every day, and electron microscopy requires that a lot of things go right. You may want to consider a (very) gently used model, a significant warranty from the manufacturer, or a new model. In contrast, if you're looking to purchase a centrifuge that doesn't need any bells or whistles, a used model will still spin just fine. You can wash out and autoclave used glass bottles just like the new ones; however, check that your new cell culture media bottles weren't last used to hold a liter of mercury solution (Person A: "Hey, everything in the incubator looks silvery and dead." Person B: "Well, did you wash your hands?" Person A: "I...I don't think that's the problem.").

2. Is this particular model prone to breaking down or does it require significant maintenance?
   a. How easy is it to get replacement parts?
   b. How easy is it to get a repair done for this machine?

---

[50] Other Author. "Buying Used Lab Equipment." The Scientist. November 12, 2001. http://www.the-scientist.com/?articles.view/articleNo/13690/title/Buying-Used-Lab-Equipment/

You have a whole internet full of reviews and websites exclusively for reviewing lab equipment (Biocompare is one such site). You also have colleagues and friends that you can talk to about the thing you are considering. There are ways to find out if the machine requires constant maintenance. If the model itself isn't reliable, then there's no need to give yourself that headache-pass on that one and see if another is available.

Alternatively, if you can find a lot of replacement parts for a good price, then something that you can easily repair if anything goes wrong could be a decent deal if the price difference between the models is significant, e.g., machine + replacement parts cost $3000 and the other model costs $7000.

3. **What are the differences between the new model and this used one?**
   a. **Do I need those new functions or not?**

Not every new function is worthwhile. For example, as of this writing, there are some new thermocyclers available where the main change from the last model is that you can turn the thermocycler on via your smartphone and receive alerts about the machine. I, however, don't care to receive alerts to my smartphone that the PCR reaction is done (because I already included that in my planning, so I should remember that it'll be done by 3pm). So should I pay $13,000 for phone alerts that I don't care about or $6000 for last year's model? Do I really need smartphone alerts for $7000? What features do you not need or care about?

4. **What is the price difference between the new and used model?**
   a. **What other things could I use that money for?**

Again, looking at thermocyclers, people have bought perfectly good used machines on eBay for less than $1000. A cheap new thermocycler can be up to $6500. Are there important consumables or other pieces of equipment that I want that I could use that $5500 savings to pay for? Economists call this the *opportunity cost*: if you decide on one course of action, that is at the sacrifice of every other action you could have taken at that time, or, alternatively, every dollar you spend on one thing is money you can't spend on anything else. What are the opportunity costs involved in buying the $6500 piece of equipment? What other things could you purchase or save for?

5. Do I really need to have this equipment in my own lab or can I work with someone else's machine?

This will be discussed later in this book, but the decision to partner with someone versus doing it yourself also applies to equipment. Are there shared resources that you can use? Some groups have shared, for example, mass spectrometers, and if you're not using that machine every day, it can be worthwhile to share the machine (and therefore share the associated costs- this is the idea behind many core facilities). It doesn't have to always be in your personal lab space- it just needs to be available when you need it.

6. Can I build an equivalent myself?
a. Can I repair the item myself?

Building your own machines may not be applicable to everyone's work, but I've built incubation slice holders from plastic cups and aquarium netting or slice 'harps' from bent stainless steel and thread. Regarding repairs, I have melted silver chloride myself and put it onto ground wires and soldered wires together. Is there something you can make for yourself that will get the job done or should you purchase the equipment? Do you need someone else for the repairs or can you do it yourself? Search the internet for what others have built themselves and determine if that's cost-effective for you.

Get multiple quotes and check out vendor promotions

No, really. It's basic, but multiple retailers can offer substantially similar equipment at different prices. Amazon sells scientific equipment, as do VWR, Thermo Fisher, and many other companies. You can also check out the different vendor promotions that are available; sometimes you get a free box of something (I planned out a series of experiments with gloves that I received for free) or maybe a ridiculous hat. The offerings can differ by what's available as well as the volume, e.g., some companies may sell the thing you want in packs of 50 when you only want one, and there's a retailer that will sell you the item in a 5-pack, which is much better for your needs.

The pro is that you can find deals on all sorts of things and see what promotions are happening for each company. The cons are that it takes time and patience, which (depending on your situation or lab manager) may be in short supply, or your institution may have cut a deal with a specific vendor and you're stuck with that vendor (unless you have a way of getting around it, and many of the vendors that have deals don't extend those deals to used equipment).

Demo equipment

"Demo" equipment is short for "demonstration," i.e., this is the machine that the salesperson took out and showed off at trade shows and product demonstrations. Generally, this is "used" and received a lot of travel-related wear and tear, like scratches on the side or the touchscreen was used a great deal, but not much usage-related wear and tear, like coffee spills or broken tubes inside of the centrifuge (or, you know, dangerous viruses if you work with clumsy people).

Generally, getting your hands on this requires that you interact with the salesperson who does product shows or knows that person. You can ask about these and what warranties, guarantees, or repairs come with the demo machine. You can often get significant savings just by choosing a new-ish demo machine. In general, you'll need to ask about demo models of the latest and greatest things or slightly older items; you won't get demos of 5-year-old models. Demo machines are also typically refurbished before they are sold to you so that most of the wear should be cosmetic, not wear that will affect your ability to do your experiments.

Refurbished equipment

To start, the term "refurbished" can describe multiple kinds of like-new equipment, and you have to ask questions to make sure that you understand exactly what was done to the equipment. Refurbished can mean the following:

- A technician has replaced and/or checked all of the parts and checked that it works;
- The internal parts have all been replaced, and the thing works like new;
- ...or it's basically a scratch-and-dent model (i.e., a machine with cosmetic problems that still does what it's supposed to, it's just not pretty).

You can get refurbished equipment from private sellers (optimal for pricing), the original manufacturer (optimal for quality), or companies that specialize in fixing up old equipment. If you have old equipment and are looking to upgrade, the original manufacturer may be interested in taking the older model off your hands in exchange for a discount, which can eliminate the headache of trying to sell the thing yourself.

The pros are that it's cheaper and functions like new or pretty close. The cons are that it's not always as cheap as the as-is used stuff and may not be the latest and greatest model.

eBay and auction sites

There are multiple websites that will let you bid on or buy used equipment.[51] These websites have a glut of lab equipment, from thermocyclers to graduated cylinders, from oven mitts to lab coats. You put in some bid before the timer expires (e.g., $10 for a used hemocytometer) and, if you put in the highest bid, then you win the item and it should be sent to you after you pay for it (and shipping, usually).

The benefits of these sites are that the items are pretty cheap, the items still work, and the sites have a wide selection of items.

The drawbacks of these sites are that you buy the equipment without seeing it, you need to keep an eye on the auction to make sure you win, the site may not always have the thing you're

---

[51] And, as you may have guessed, you can sell your old lab equipment on these sites (provided that you're allowed to do so, as some universities/nonprofits have their own auctioning processes).

looking for in stock, the seller may not send the item (though you have some recourse if they do not), and the equipment can be quite old.

I'll expand on each of these points.

- The items you can find are cheap. It's used equipment that is in a niche market, so you have some room to negotiate (especially if you see that it's been on the site for some time without being sold). You can save a great deal of money on equipment; some new items are over $10,000 and I've seen older used models on eBay for $1000; 90% savings is nothing to sneer at if the discounted equipment will still accomplish your goals.
- The items still work. If you're getting a used item from the original manufacturer or a third party with a trustworthy warranty, then the machine should still do what it's supposed to. It's like buying used cars: I can keep upgrading used cars every couple of years and spend less than $20,000 altogether in a 20-year period, while a new car can cost over $40,000. If I have to replace this item in 7 years, I'm still likely paying less than if I bought one new car. However, if you're getting the item from an unknown seller (as in you can't find a customer service line to bother them), then you're buying the item as-is and it may or may not work.
- The sites do have a wide selection of items. You can find many different pieces of equipment from all sorts of different fields on these sites and it's likely they'll have something for your field. You can discover how many

different manufacturers there are (or were) for a specific piece of equipment that you need.

- Since these are currently exclusively online, you buy the equipment without getting to see it first. As such, this is best done if you have access to an electronics shop (in your department) or a cheaper repair service so that you can have someone evaluate the machine when it comes in. If you've saved $5000 on the device and paid $500 for the repair, you've still come out ahead and have a machine that works.
- You have to watch the auction. If you are beaten by one cent, you still lose the auction. There are automated alerts and systems that you can use or you can check it once a day (until the end of the auction). This can take up some more of your precious time and energy.
- If you need something specific, you are at the mercy of whatever someone happens to be selling on the site; I've never seen a scientific equipment site that would let you put up an "I want this" ad to get offers (though that's a great opportunity if you find it). Unless you're looking for very common equipment, this likely can't be the only place you look for supplies.

Before using a site like this, check what you can do if the seller doesn't actually send the item to you after you've bought it. Does the seller have a website or is the shipping guaranteed by the company that hosts the auction? The risk is low for sellers that are relatively well-known, but it is not one that can be completely unconsidered.

Finally, the age of the equipment and how well it was maintained can also vary considerably between sellers, which means that you should have someone (or yourself, if you're handy) go through the machine after you receive it to make sure that it will meet your needs, i.e., the machine will work for the experiment.

In-person auctions

Many universities and other entities will host their own auctions of old equipment. You can see the item (though probably not examine it thoroughly) and you know that they'll ship it (I mean, it's right in front of you), but the benefits and drawbacks are mostly the same as those of online auctions. You may also have to sit through auctions for a number of items that you aren't interested in. If possible, inquire with the office about whether you can just have the item sent to your lab (if you're in that university) for free.

Free stuff (aka freebies)

There are multiple ways to get equipment for free, especially if you're in an academic lab:

- Specific governmental or nonprofit associations that give away equipment to academics
- Picking up items through university auction or recycling programs
- Going through the trash
- Convincing the dean or chair to buy a piece of equipment for everyone to share
- Gifts from friends

Typically, the governmental or nonprofit associations that give away equipment are more involved in niche fields (not molecular biology, at any rate). University auction or recycling programs can be an excellent source of materials; a less formal version of this is asking a neighbor if there's anything they're getting rid of (I once got some rubidium this way).

If a lab head is leaving and not taking all the equipment or consumables with them, you can ask for them or ask whoever is in charge of disbursement (different institutions have specific policies to follow) if you can have some of the items. Always ask and always be polite- they are doing you a favor.[52]

Going through the trash is risky- there's a significant risk of being penny wise and pound foolish. You can pick up an expensive machine that was left out to be picked up, but it may not meet your needs (or require a lot of time to make it work the way you want it to). Unless you know and ask the former owner, you don't know what was wrong with it or what maintenance it needs, unless they were throwing out the owner's manual as well.

Vendors sometimes offer free stuff related to equipment, but not typically the equipment itself (glass bottles are a rare exception). Don't badger the sales guy for a free $30,000 microscope- you're not getting one. You can, however, ask him to give you some extra solutions or an additional electrode for free if you're buying a pH meter, for example.

---

[52] I once saw a lab that was raided by colleagues before the group finished moving out, which prevented them from finishing their experiments because others had taken all of their equipment already. There is no polite way to describe this behavior, so don't do it.

Stuff that doesn't cost *money*

There is a difference between free and things that don't cost money; you can ask different groups (companies, other researchers) for equipment or consumables, and they may share the item but may also include different requirements about what you do with the equipment or consumable item. For example, a company can state in a contract that they get to publish any of your data within 6 months of completing the project if you don't publish it yourself. Alternatively, they may state that you can use the item but in exchange for a confidential report on everything that went right or wrong with the machine.

If you go this route, you must read the contract and be certain of what the real cost is for getting the item, even if money isn't involved.

Do it yourself (DIY)

Making equipment yourself has a proud history in science; after all, someone had to be the first to invent the thing. The internet abounds with tutorials about how to make different things for the lab and in general.[53] If you like working with your hands, you can make some interesting things (whether they save you money depends on the source(s) of materials that you use).[54]

Making things yourself can also add some individuality to a lab; I have seen agarose gel boxes with wood paneling, cells fed with medical tubing lines, and wooden Faraday cages.

---

[53] Have you ever wanted to make a solar-powered phone charger for your backyard? The internet has you covered.

[54] Reclaimed pine wood? Cheap. Cherry wood? Expensive. Old soda pop cans? Cheap. Aluminum alloy bars? Expensive.

The pros are that it can be really cheap and, if your problem is unique, the custom solution can be a better fix than looking at what is available outside of the lab.

The cons are that if it breaks, you're the only one who can fix it (at least in your lab) and that, to build the stuff, you should know what you're doing. If you don't know what you're doing, it may not solve your problem![55]

Maintenance

Doing your own maintenance actually has a number of benefits. If something breaks, you can usually fix it faster than it will take to schedule a service call (where a repairman comes out to see you) and meet the repairman, who may have the right part with them (or if they don't, schedule a second service call for them to come back with the right part), and then fix the machine. Doing regular maintenance also decreases the number of catastrophic failures that the machine will experience during its working life.

If you know what the machine is capable of doing, you can design different experiments that take advantage of the full functionality of the machine, i.e., there may be additional functions that the machine has that weren't described to you initially. For example, I've seen a household blender used to turn chemical chunks into fine powders, a PCR thermocycler with cooling blocks used as a

---

[55] I once knew a guy who developed a program to play a certain card game. I invited him to play the card game with me in real life. He said, "No, I have no idea how to play." I replied: "...then how do you know that your program worked?" I did not use any of his programs.

small 4°C refrigerator, and a drip coffee pot used to prepare ramen noodles.

This can be as simple as sitting with the owner's manual, reading it, and ordering the right parts. Doing this can save you a bit of money and time, but it does cost you time and patience. Although, if you're like me, it can also be a lot of fun.

Methods of Saving on Equipment Pro/Con Table	
Multiple quotes	
PRO	CON
Can find better prices, volumes	Takes time and patience
May find exactly what you want due to differences in availability between vendors	May be stuck with a specific vendor and this doesn't apply
Demo equipment	
PRO	CON
Cost savings	Travel wear and tear
No usage-related wear and tear	Needs relationship with salesperson
Manufacturer warranties, guarantees	
Refurbished equipment	
PRO	CON
Cheaper	Used
Works like new	Not as cheap as other options
Manufacturer warranties, guarantees	

Methods of Saving on Equipment Pro/Con Table	
Auction sites	
PRO	CON
Significant savings (up to 50-90%)	You're buying it without seeing it or knowing its history
Wide selection	May not have what you want in stock
	Age and quality can vary greatly
Free	
PRO	CON
Can't get cheaper than free	Selection is drastically limited
	All the problems and maintenance issues are yours to deal with; companies generally don't want to help you with equipment you didn't buy
Built it yourself	
PRO	CON
Cheaper than many other options	If it breaks, there's not really anyone else you can call to fix it
May solve the problem better as a custom solution than commercially-available options	You really have to know what you're doing!

Methods of Saving on Equipment Pro/Con Table	
Do your own maintenance	
<u>PRO</u>	<u>CON</u>
Significant savings	Time spent on maintenance could be spent on other things
Quicker to fix problems; don't have to wait for someone else's schedule to open up	
Knowing your equipment better may open up alternative uses for the same machine	

Chapter 8

Score deals on consumables

We owe modern life science to plastic.

Get multiple quotes and check out vendor promotions

This is similar to the same advice for equipment. Many vendors sell lab consumables, and there are deals or promotions every so often, especially if you're willing to try something new. Amazon, VWR, Thermo Fisher, and many other companies sell the relevant consumables at various prices and volumes. I once received enough free gloves, 96-well plates, and DNA extraction kits for a summer's worth of experiments from a vendor promotion, so there are rewards for the patient.

If you're dealing with different sales reps, it's okay to point out the different quotes you have received (but don't lie about it). They are empowered to make different deals, and if you have developed a good relationship (i.e., not antagonistic all the time), the rep may come to you when they have a deal they know will be useful for you. Speaking of which...

Work with the sales reps

The job of the sales rep is to sell things, not to give you things for free. They, however, have more information than you do about what deals are going on, what things need to be sold, what things can be given as bonuses, and more. If you have specific vendors that you would like to continue working with, then just be open and transparent about your needs and when things are available-they will understand if you point out that x is your budget and you'll hear back about money in y months. If they come by, just make sure to ask them what they can do for you with this latest order on your mind. You can always look up competitor prices and start negotiating.

The main idea is that they are people too, so work on having an amicable relationship and that can pay dividends later (like most human relationships).

Do yourself and the salesperson a favor: come to them when you want something specific, have money to spend, and want a deal. They can, and in my experience will, come through for you if you understand how their job works. Additionally, the sales reps from specialized companies typically double as troubleshooting experts and are available to answer a few questions when they come around your lab to check up on you. It is absolutely okay to ask them to come by and help you use the machine or device that they sold to you.

Free stuff

Vendors may also have freebies that you can pick up, either from their website (which waxes and wanes in popularity by how the economy is doing, generally) or trade shows (this is much more common for consumables than for equipment). There are three main benefits of going to these trade shows:

1. Seeing your sales reps face-to-face can help develop your relationships with them.
2. They offer all sorts of lab-related goodies that can come in useful.

3. They usually offer lunch, so a bunch of students and postdocs that forgot to bring lunch can score some pizza or sandwiches.

The trade show goodies may not be directly relevant to your lab, but they can be useful. I have received (for free):

- So many flash (USB) drives that I haven't bought a new one since 2002, including one shaped like a credit card and another shaped like a scientist;
- Multiple coffee mugs;
- Many pens and highlighters that fill those coffee mugs;
- Calendars;
- pH meter solutions;
- More notepads and bookmarks than you could hope for;
- Gift cards for coffee shops;
- Consumables (e.g., gloves, plastic tubes, and cell culture flasks);
- A laser pointer with my name laser-etched onto the side;
- Troubleshooting guides for relevant techniques;
- Tools to help me adjust microscope lenses;
- Free samples of serum for cell culture;
- Free samples of primary and secondary antibodies for western blotting (which were life-savers at the time);
- Glass reagent bottles;
- Capsules to make LB broth, which led TSA to check my bags (sorry, guys);
- A stress ball shaped like a coffee mug;
- Free samples of x-ray film and chemiluminescent development kits;

- Molecular biology kit samples that I used to run multiple experiments (that were later published);
- One of my favorite permanent markers (an ultrafine-tip permanent marker);
- Some very nice shirts;
- And some truly awful shirts that are best worn while doing work that would stain your clothes.

University recycling/sharing programs can be a great source of unopened consumables or consumables that generally don't go bad, like 95% ethanol. They often won't charge for you to get the items. Similar to the last chapter's advice on equipment, you can also ask (amicably) departing labs about some of this stuff, since a lot of this stuff is cheap and costs more to ship than to just buy new bottles when you get to the new spot.

Ask your local core facilities if they are getting rid of anything- I once got a 2-year supply of cell culture serum that had just sat in their freezer (I tested a batch first- it worked just fine). If a lab doesn't want to continue performing a technique that you're interested in, you can work out a deal with them for their consumables related to that technique (before anything goes bad, of course).

Don't even consider going through the trash for consumables- if the other lab is throwing out the consumables, something happened, especially if that consumable doesn't go bad. A true story- a "perfectly good bag" of pipette tips happened to have been covered with moldy cell culture, which dried up and wasn't really visible. The new cell culture just happened to have mold problems.

Aliquot your consumables (if applicable)

Aliquoting is the practice of splitting liquids into multiple clean containers to minimize freeze-thaw cycles or contamination (which I described in greater detail in *Basic Molecular Protocols in Neuroscience: Tips, Tricks, and Pitfalls*). For example, I have aliquoted antibodies, cell culture sera, quantitative PCR master mix, primer stocks, ion channel blockers in solution, and free proteins in solution. Aliquoting may not be what you think of in terms of saving money, but part of saving money is not paying as much up front and part of it is not paying as much over time. Aliquoting can decrease the costs paid over time.

If anything you're working with can be damaged by being taken out of the freezer too many times or can be contaminated and ruin your whole month, it's a good idea to aliquot it. That way, if something appears to be damaged, you can get a relatively fresh batch out of the freezer and start over. If someone contaminated your primer stock, you can pick another stock out and start over without ordering the primer set again; if someone contaminates a common, shared stock solution, it can take a while to figure out where the problem is really coming from.

Use reusable options

While not always feasible, there are times when non-consumable options can help you save additional money. For example, using human cells in culture requires disposable pipettes to comply with health standards and applicable law in some areas. Well, if you're using non-human cells, are you able to use autoclaved glass or plastic pipettes to distribute your liquids?

Perhaps you are grinding tissue samples. There are disposable pestles that can be purchased, but would a washable permanent pestle work better on these samples? If you're willing to clean something properly, then you can save money by not needing to purchase a consumable item in the first place.

Bulk options

Bulk options can include buying items in bulk, but it also includes making things in bulk. Why does this save you any money?

Buying things in bulk saves money because less of the cost needs to go toward packaging and the supplier wants to get rid of product, so offering a deal on 50 kilograms of product moves more of it out the door and benefits the supplier.

Making a commonly-used solution in bulk saves money by decreasing the amount of times that possible errors can occur and how much of the material is wasted due to those errors.

For example, let's say that you're making agarose gels. You only need 25 ml of your TAE solution per gel. You decide to make 15 liters of the solution in bulk (15,000 ml). You make the solution once and spill 10 mg of powder on your benchtop. That's enough liquid for 600 gels. By making it in bulk, you've spilled and lost 10 mg of powder. If you spilled that amount for all 600 gels, that results in 6 grams of the powder that's gone to waste and no one can use; depending on the size of the bottle, did you lose ¼ or more of the entire bottle? TAE doesn't really go bad, so you can make it in bulk. Are there other solutions or things you can prepare ahead of time in bulk?

Any time you decrease the possibility of experimental error, you win. Making common items in bulk is one way of doing just that.

Make your own stuff

While it's not always possible (you can't make your own disposable gloves, for example),[56] you can save some money by making your own reagents. For example, it is cheaper to purchase the original chemicals and make your own stock solutions for chemiluminescence (and may also be more effective). You can also make your own loading buffers (e.g., for agarose gels) much more cheaply than buying the ready-made stuff.

If you're living and working in a money-scarce environment, knowing how to make do with the original chemicals to make consumable reagents can be a life-saver. If you have plenty of money, then run the calculations and decide what the time is worth versus the cost of buying the consumable reagent over and over. If you're using the reagent on a regular basis, the savings can be impressive (e.g., $80 for 6 months to a year instead of $80 per month).

---

[56] At least, I typically can't. Your lab may have its own fabrication machinery and a ridiculous amount of latex, for reasons you likely have to explain to others.

Methods of Saving on Consumables Pro/Con Table	
Multiple quotes	
PRO	CON
Can find better prices, volumes	Takes time and patience
May find exactly what you want due to differences in availability between vendors	May be stuck with a specific vendor and this doesn't apply
Sales rep relationships	
PRO	CON
Cost savings	Takes time
A friendly face may be welcome	Can irritate you when you don't have any money and they still visit
Manufacturer warranties, guarantees	
Free stuff (including trade shows)	
PRO	CON
Can't get cheaper than free	May be used
Can get some fun surprises	Selection is drastically limited
Can be a fun way to spend some time	Time at trade shows is not running experiments!

Methods of Saving on Consumables Pro/Con Table	
Aliquoting	
PRO	CON
Makes life easier if you have to throw out a solution	Takes time; depending on how much you're aliquoting, the time taken can be significant
Can save money over time or mitigate the cost of training new people	Need to prepare the other plastic tubes or containers ahead of time
Reusable options	
PRO	CON
Cheaper over time	May be more expensive up front
May be more effective	More time-consuming (cleaning vs. just throwing the item away)
Buy in bulk	
PRO	CON
Decreases the per-unit cost significantly	If it turns out to be a bad deal, you've got a lot of it to go through first
Can decrease experimenter error	

Methods of Saving on Consumables Pro/Con Table	
Make your own solutions	
PRO	CON
Significant savings	You have to know what you're doing!
May solve the problem better than commercially-available options	Can take a lot of up-front time
Once it's made, you just go get it immediately	Can cost more initially to get esoteric chemicals in the amounts needed to make bulk chemicals
If made in bulk, can decrease disparity between different researchers doing the same thing	

Chapter 9

Budgeting, technique optimization, and collaborations

"If you don't know where you're going, you'll end up someplace else." -Yogi Berra

"Give me six hours to chop down a tree and I will spend the first four sharpening the axe." -Abraham Lincoln

Why you should plan and run projections on each experiment/project (budgeting is important)

You probably run projections all the time. Have you ever calculated whether it was worthwhile to buy a house versus renting one? Have you ever calculated whether you could afford a new car by looking at how the cost per month would affect your budget? Have you ever made a budget? If so, you've made projections.

There is a great benefit to making sure that your experiments are efficient, which may be more visible in some types of experiments than others. For example, I used to work with a lot of 96-well plates. Some of these plates were specialized for use in qPCR machines and only in specific qPCR machines. In that case, these plates were more expensive than most plain 96-well plates. Therefore, I didn't use any plates unless at least 95 wells were being used for something.[57] This decreased my overall costs for different experiments.

Another example of experimental efficiency: I once had a drug that I used in pharmacology experiments that could survive only one freeze-thaw cycle. I determined the desired drug concentration and what volume I wanted to use the drug in, and then made aliquots of just that right amount. One tube = one experiment, without any wasted drug. This also made work go faster, since I had just the right amount (so just toss the whole thing in at once in a given volume of liquid, no extra measuring) and didn't need to thaw more than one tube. I didn't have the drug start to lose effectiveness due to extra freeze-thaw cycles nor did I have to buy additional amounts of the drug because some of it was wasted.

You may be thinking that there is more to life than squeezing out every last ounce of efficiency in the day. You would be right- I thoroughly planned and squeezed the efficiency out of my work so that I could leave work after 5:30 so that I could see more of my kids and help put them to bed every night. Having a relaxed evening was more important to me than having an unplanned day. Being efficient generally isn't done for its own sake; planning gives you the opportunity to do the things you want to do when you want to do them.

Making projections for the cost of your experiments is part of the process of thoroughly planning the experiment. This can be exact, or you can do this with back-of-the-napkin figures (i.e., rounded for the sake of time and ease of calculation). The point is to

---

[57] Data are data; the plate doesn't care if you're running parts of 3 different experiments on the same plate or just 1. In that case, you simply need to keep good notes and keep track of what is where. You can even make notes in the software that runs the machine.

understand what something should cost so that you understand if that particular experiment will bankrupt you or not. Let's run through an example scenario to illustrate what I mean.

- You have a budget of $5000 for this experiment. Facilities and overhead (i.e., lab space, electricity, and running water) are paid for separately and not part of this particular budget.
- You already have mice for this experiment; you don't have to pay for the animals' shipping, just their care ($0.80 per day).
- You want to train a total of 20 mice (10 treated mice, 10 controls) to run through a maze and measure a specific protein in their blood once per month.
- You will use ELISA to measure the protein because you already have all the equipment and experience needed to analyze ELISAs.
- Your experiment will ideally take 6 months to make sure the animals are really well-trained.
- You don't currently have the needles, syringes, or blood collection tubes for this experiment.
- You already have the maze for this experiment and the associated equipment (cameras, computers, etc.).

Let's see what this looks like in a table, which might be easier to understand:

Item	Cost
20 mice, 2 groups of 10	$0.80/day per mouse * 180 days = $144 per mouse for 6 months; $2880 total
ELISA plate for 128 reactions (120 data points, 8 controls)	$490
20-gauge needles with disposable 1-ml syringes (need 120)	$30 per 100, need 120 = $60
Blood collection tubes (need 120)	$30 per 100, need 120 = $60
Running Total	**$3490**

In this setup, you would collect the blood and run one single ELISA with everything on it. Is that feasible? Does the protein break down in the -80 °C freezer? Should this experiment be planned differently? You can, if you want, add another 3 plates total if necessary, or add additional mice (7 max, since while you already have the tubes and syringes necessary, you'd need another ELISA plate). Are you paying a technician for this or are you doing the experiment? If you need someone else to do the work, the cost of staffing is far too significant to be captured in a $5000 budget,[58] and so you need to either vastly increase the budget, find

---

[58] Your paycheck is one cost, each of your benefits costs extra, the insurance for having someone on staff is another cost (e.g., unemployment insurance held by the employer), workers' compensation,

a free worker (internships and the like), or do it yourself. Are there better deals on the materials right now?

The objective is to run the best experiment that you can within your given constraints. If that constraint is time, then plan out the day and how things should proceed. If that constraint is money, then plan out how to best approach the problem within your budget.

In contrast, let's say you just planned the experiment with a better number for your purposes (32 mice, 4 groups of 8) and didn't order the ELISA plates until the end.

Medicare/Medicaid/Social Security taxes, Federal and state unemployment taxes, your tuition (if applicable) is another cost, and the list goes on.

Item	Cost
32 mice	$0.80/day per mouse * 180 days = $144 per mouse for 6 months; $4608 total
20-gauge needles with disposable 1-ml syringes (need 192)	$30 per 100, need 192 = $60
Blood collection tubes (need 192)	$30 per 100, need 192 =$60
Running Total	**$4728**
ELISA plate for 128 reactions (can only get 120 data points plus 8 controls on a plate, instead of the 192 blood samples you collected)	$490
Total	**$5218**

You may have completed the experiment, but waiting until the end to buy the ELISA plates left you with 192 vials of blood and...no data, since buying even one plate went over the budget. If your lab is in real danger of running out of money, that means 6 months of your life has gone by and you don't have a completed experiment to show for it. That's a tough spot to be in and explain.

Okay, but what if I'm not in dire financial straits?

Most of you are likely not under the strict budget that I was. If you have a budget for the project but it's not all that strict, then having a plan gives you greater flexibility. For example, let's say you were really efficient and saved $500 on your last experiment. You're writing up the results and you're pretty proud of yourself.

Then a blockbuster paper comes out that points out how one specific protein could really affect your experiment.

The antibody that could make your experiment even better is $400, and the experiment would only cost you $80 to re-run for that specific protein.

Look at that! The $500 you saved in one spot can be now used to make your paper better if you wanted to pursue that option. Or you could just get the antibody and use it in a later experiment.

Or you could do something that led to one of my favorite times of year. You see, there were grants that had to be spent by a certain date. So there would be a day where I was told to spend so much by a certain date. Because I'd been frugal up to that point, I could plan out different experiments and get deals on buying certain items in bulk. I could plan collaborations and get my share of some items for those collaborations. I did plan a year's worth of experiments at one point and acquired what I needed to run those experiments and get things to share with my collaborators.

If the budget had been higher, we could have pitched in to help get some $200,000 device (or more expensive).

Or, if you're in dire straits, you can try to get the items that you need at a bare minimum to keep things running (for example, enough cell culture media to swim in if your lab runs exclusively on cell lines). This can give you more time to keep working while you apply for grants and sources of funding.

Running a lab is, in its own way, like running a fresh startup: if you run out of money, you're done. So the goal of this section is to encourage you to be responsible with what you have so that you don't find yourself running out of money and in a bad situation.

Optimize your techniques

"Working hard and working smart can sometimes be two different things." -Byron Dorgan

I am a fan of optimizing techniques in order to save time and money. Not every technique can be optimized to the point where it's considered cheap; we are in the age of sequencing individual patients' genomes, but only by purchasing them in bulk do we get them to cost less than $1000 each.

However, the cost of human genome sequencing shouldn't stop you from examining whether you can optimize any of your techniques. Let's examine a few reasons why you might decide to thoroughly examine and re-evaluate your own ways of doing things.

Efficiency

The first thought is that it can increase your efficiency. You will need to determine which steps of the technique are essential to its success (e.g., autoclaving everything), and which ones are kept due to old biases (if the technique was developed from another technique), lab superstitions (e.g., orange socks decorated with ninjas are required for the microscope to work), and because no one else bothered to see if the results were any different once this particular step was omitted.

If you determine which steps to cut out, perhaps it only saves you 5 minutes of your life. Maybe you're thinking, "Eh, five minutes isn't worth all that fuss." Does it save you any money to omit that step? If so, you're now saving some money and time. What if, as happened in my own experience, you save 5 minutes on something you do fairly often? Let's say you save 5 minutes on something you have to do 8 times in a day (extracting RNA or proteins from samples, for example). You now have an additional 40 minutes to do other things, like sit and actually enjoy your lunch.

As another example, I once discovered that my RNA yield increased in one RNA extraction protocol when I omitted one standard step without any change in my overall RNA purity. That meant that I needed to extract RNA from fewer samples because that increased yield stretched further to additional experiments. I saved the time of getting the samples, making the solutions, performing the extractions, and testing the purity of the extracted samples that I would have needed to collect if I continued using the old protocol. I effectively saved multiple days' worth of work by paying attention to the process itself.

Just like with your wallet, small savings of time can add up.

Cost-effectiveness

Related to the example above, let's say that omitting the unnecessary step means that you no longer have to purchase a particular reagent. You now can either find an alternative use for that reagent, barter it away, or just enjoy not having to purchase it any more.

Alternatively, I read papers that found generalized ways to prepare samples, e.g., collecting the RNA, DNA, and protein from a single sample, obviating the need to collect 3 samples in 3 different ways (separate RNA, DNA, and protein protocols) to collect 3 different things- this can save quite a bit of money if your lab has one way that they collect all the pieces that you need.

Quality

Interestingly, you can sometimes get superior results by changing the way things are done. For example, one technique that I optimized gave better results (for both yield and purity) when I did not heat the samples at the end; it turned out that the extra step denatured some of my extracted material and washed away some of the still-good material. You may find that there are some variable steps that can change your yields or function if you changed one reagent or two. In another technique, combining certain reagents increased the yield better than using either reagent separately, which cut down on the number of preparations required per sample.

If increasing the quality of the preparation can decrease the amount of times you have to perform the prep, you have saved both money and time. If you only have to do it once, then it's cheaper to do it right the first time instead of needing multiple preps.

Collaborate (partnership), buy, or DIY?

"Coming together is a beginning; keeping together is progress; working together is success." -Henry Ford

When it comes to your lab and equipment, you will come up to the decision that many small businesses do: for a new function (technique, machine, idea, etc.), do you collaborate (work with a partner or partner lab), pay for something that will accomplish the function for you (send out to an outside company, for example, or hire a new postdoc for just this function), or will you or someone already in your lab learn how to do it (i.e., keep it in-house)?

Each strategy has its own pros and cons, and the choice you make will depend on you and your vision for how the lab will be run.

Collaboration

Any collaboration or partnership requires a certain degree of trust. Collaborations are easiest between friends, so long as an agreed-upon set of rules is in place- you want a previously agreed-upon set of rules to avoid any heartbreak, hurt feelings, or surprises. If you don't have the money to hire a new person or send out the materials, collaboration can have many benefits. If the collaborator is in your institution/company, then you have similar goals and incentives, so the pressures you are under are likely similar.

If the collaborator is in another institution, then there are likely similar goals and incentives, even if they are a competitor; however, collaborating with a competitor can present its own challenges and headaches. If possible, it is better to work with someone who is an expert in the technique but not necessarily someone you compete with. You can try to get a work-for-hire contract to have a competitor teach you how to do x, y, or z, though not everyone will agree to do so, even in exchange for cash. Alternatively, you can use services that you are able to provide as a source of funding for the lab, but a work-for-hire or service-for-hire is different from a collaboration and requires a different mindset (it's a business agreement and business setup in the middle of a non-business environment, typically).

The benefits of collaboration are obvious: you get a (possibly long-term) partner who adds the function to your existing work at typically no monetary cost to you. The partner thinks differently and approaches the work differently, which can enhance your own thinking and make the resulting product much richer than it would have been if you'd done it alone. The work can lead down multiple avenues for future experiments, including scenarios where the other lab needs *your* input about how to proceed. Collaborations have great potential to be very fruitful- I have seen collaborations that kept giving for over 15 years, some for over 20 years.

The drawbacks of collaboration are that you need to give them something in return, whether that's space on the paper (some labs are more protective of this than others), intellectual property, or even splitting grant funding. Depending on your situation, that may be acceptable, remarkably cheap, or completely unacceptable. You will need to negotiate before the collaboration starts and possibly after; again, previously agreed-upon rules can prevent a lot of heartbreak and bitterness.

Regarding new machines, a "collaboration" typically looks like either working with a core facility (who will likely charge you for the time you spend on the machine) or splitting the costs with multiple people to buy the new machine (e.g., 10 labs putting in $100,000 each for a $1 million lab-ready mass spectrometer).

The benefits are that you have access to the machine, you did not need to pay full price, and someone else does the maintenance tasks.

The drawbacks are that you might be charged for the time, you need to share time on the machine, and if there isn't a use-the-machine-this-way procedure in place you may run into problems created by someone who is unaffiliated with you or your lab.[59]

Collaborations are easiest to set up with people you already know; beginning a collaboration online is growing in popularity, but it can be quite difficult to check up on progress (someone can just ignore the email, after all) and requires that you set up multiple safeguards in case the person you only know via email is unreliable.

I set up multiple collaborations by simply knocking on doors and discussing my work with other people. Persistence is the key if you want to start working with a new person (or lab).

Outsourcing or hiring a new person

Outsourcing can be more expensive than the other two options in terms of monetary costs, but you don't have to share any intellectual property or publication rights with another lab.

Outsourcing looks like, for example, sending tissue samples to a sequencing laboratory that offers its services on the open market.

---

[59] "Oh, yeah, sorry, Jerry replaced all the default settings on the machine, changed out a few filter cubes, and adjusted the laser's wavelength...it looks like he took absolutely no notes, just left for vacation, and won't be back until next week."

You will own the data, but they are going to charge more for the same service (after all, they have to cover their overhead and technician salaries). If your lab is small or you don't want to grow into this particular technique, then outsourcing can work well.

The benefits are that there will typically be contractual obligations and guarantees that the work will be done and done well. You can get the data quickly without needing to get yourself or anyone else up to speed on how to collect or analyze such data.

The drawbacks are the cost and the time that you will need to wait for the results, which can vary by your chosen contractor. Maybe you get your results in a week, or perhaps you will need to wait a month.

Another type of outsourcing is hiring a new person who can do the technique or function that you want done. Hiring a new person for the exact function can work out well, but there are a few considerations that you must pay attention to.

First, since you don't know how to do the technique, you are not able to critique their work like an expert could (at least initially). You will need to find others to support the person (if they encounter something they haven't solved before) and teach both of you more about the technique. This will take time and the person may not be as productive immediately as you both imagined.

Second, you must get along well with the new hire. While this is at least somewhat true for all hires, this person can do something that you can't do and can't fully critique, which means that they are in a position of significant trust.

Specifically, you have to trust that they are doing the work in a way that is standard (or at least not suspicious) and that they are designing the experiments properly. If you can't trust them (or can't trust anyone in that position, really, because you can't evaluate their work properly), then the new hire has the potential to be a catastrophic failure since morale issues spread quickly through labs, the money is gone, and both you and the hire are frustrated.

Third, if you had to order new equipment for this, do you know that you ordered the correct equipment? Will the new person have the ability to order (or at least request) additional equipment that they may need? Will you wait for them to arrive before you order the equipment, which will delay their productivity but ensure that the correct equipment comes in?

Fourth, because of all the costs associated with hiring a new person, this can be more expensive than other options, especially if you find you only want this technique for one experiment.

The potential benefits of hiring a new person is that you decrease the amount of time required to get up to speed on a particular technique. The new person can also train others in the technique, which will help promote the technique and enable subsequent hires (or trainees) to perform the technique.

The potential drawbacks include the cost, that the new hire will know more than you about the technique in question, and the costs of additional equipment (including replacing obsolete or ineffective equipment).

Do it yourself

You can also decide that you're going to develop the capability in your own lab. This is a little different than hiring a new person who will be able to walk in and perform the technique or function you want- this is taking someone you already have off of their current work and training them to start on the new technique/function or you taking more of your own time to learn this new thing.

One possible way of doing this is to start a collaboration with this in mind; you can trade technicians for a period of time so that their technician learns your skillset and your technician learns the other party's skillset. Afterward, you ideally have someone who can perform the work and train others to perform the work. This can also be acquired through a sabbatical at another laboratory.

The cost of your time is something that should be considered. You will likely be busy with coursework (if you're a student), teaching courses, meetings, planning out your own experiments, applying for grants, running other experiments, preparing for experiments (making solutions and the like), performing public service or volunteer work (new faculty), and other considerations. Can you fit in a new field or new technique? How much reading, training, and hard-won experience will be required to start to get an idea of how to do this? Specifically, is this a technique that works with well-established recipes and instructions (e.g., plain PCR) or is this a technique where the really good people are measured by whether they have at least 10 years of experience (e.g., *in vivo* electrophysiology)?

Importantly, how much support will you have to get this started, in terms of money, people, mentors, and time?

The benefit of doing it yourself is that you now have this capability and don't have to trade anything for it. You can teach it to others and can critique their work as an expert.

The drawback is that it takes time and effort to learn something new. If you have your technician take the time to learn this new thing (or you do), there is a trade-off between what they (or you) could have done with that time instead of learning the new thing. Additionally, if this technique requires the purchase of additional equipment, then the equipment costs will have to be factored in.

Chapter 10

Intellectual property (IP) matters: books and inventions

Intellectual property is the engine of modern economies and what you create needs to be protected.

Intellectual property (IP) is a growing concern at the modern university, especially after the Bayh-Dole Act of 1980 created the modern institution of technology transfer (likely an office at your university or a department at your company, if the company is large enough). IP is property that is the result of someone's creativity (can also be called "creations of the mind"), like a painting, an invention, or a book like this one. As a researcher, there are a few IP classes that pertain to your work:

1. Utility patents (inventions)
2. Design patents (some sort of aesthetic design feature in an invention or device)
3. Plant patents
4. Copyright (books, works of art, etc.)
5. Trade secrets (this is mostly a concern for companies)

The reason that you should care about IP is because you have the potential to benefit from it, in terms of money, fame, tenure (some departments consider patents when determining whether to grant tenure), and (or) the satisfaction of your invention going out into the world and helping people. A book shows that you have some expertise in the field (and is an accomplishment in itself). A patent clearly shows expertise in your field (you used your background to bring a new idea into the world as an invention). A painting can be sold or be something that you get known for. If your invention is licensed by your employer, you may get some of that revenue- if you work for a university as a professor or for an unusually permissive company. If you want to start a company around your invention, the startup can license the invention from your employer and you can participate in growing a company from scratch, which has a very high failure rate but can be a wonderful experience.

Your employment contract and employer usually have standard practices regarding who owns what, but you can always ask if your creation is owned by you or by another. For example, if your employer shares some of the invention revenue with you, then you are paid a portion of any licensing fees that go to the university. Most companies that employ engineers do not share any revenue with the inventors. Different employers have different rules- here are some possibilities:

- Employer owns all IP generated by the employee as part of the employment contract;
- Employer owns all IP generated by the employee *that is related to their job*;
- All professors, researchers, and staff give up their IP to their employer, *but students do not*;
- Anything you make on your own time and can clearly show wasn't done at work is yours.

One big issue is finding the time to create IP. If it's a book, then your employer may have rules regarding how much time you can spend on that book if you're on break, for example. If it's an invention that you create in their lab space, there are likely no real ways to take ownership of that invention yourself. There are benefits and freedoms in making your own IP; you can write the course textbook you've always wanted or answer odd questions that no grant committee would like (e.g., "How has pneumonia affected European history?"), and make some money while doing it.

Who should I check with regarding IP ownership?

Are you at a company? What is written in your employment contracts? What is in the standard practice guide? Is there a technology transfer office? Does HR (human resources) handle some aspects of this? You may have signed away all your rights to any IP as part of getting a job. Make sure to look at the contract and not just take someone's word for it.

Are you at a university? Then check with your commercialization or tech transfer office. Some universities own only staff, faculty and postdoc inventions, while others own graduate student inventions too (that's in the fine print of any hiring contract you signed). Typically, the inventions of undergraduate students belong to those undergraduate students.

You should check the relevant paperwork; some places only own the inventions related to your job (so if you study laser physics but write a children's book about kittens, the book is your property, unless the title is *Meowzers and the Adorable Argon Laser*), but others own any intellectual property that you make while working for them.

If you are not at a university any more, then still check the university guides and your own employer's policies. Some universities include clauses that if you invent something related to the work you did for them, even if you are out of that university, then they still own it, even years later.

Bottom line: you can't assume that you own what you invent until you've made sure that no one else should (properly) try to claim it.

Types of IP

Patents

We will cover 3 types of patents here: utility, design, and plant patents. These are distinctions made in the US regarding different types of patent protection. If you invent something that your employer owns, then your employer will likely retain the services of patent attorneys. If you invent something and you own it, I highly recommend retaining the services of your own patent attorney; many people fail the patent bar (the exam that allows someone to work with the US Patent and Trademark Office (USPTO) regarding patents) each year (it has a less-than-50% pass rate and most of these test-takers are engineers or have PhDs), which means that it's difficult to navigate around all the USPTO's rules and regulations as a novice, and an experienced attorney can help identify where you might be thinking too narrowly about your invention.

Utility patents

When someone says, "patent," you're likely thinking of a utility patent. A utility patent protects an invention that is a machine, process (e.g., the process of making a specific heart medication from basic chemicals), manufactured product, or compound. A patent is a contract where you get a government-provided (and legally-enforced) monopoly on that invention for a given period of time, after which anyone can use it (or at least the information is available to the public). For some inventions, a patent is mandatory to protect your IP; for example, a new drug for treating heart disease would need to go through government examination (like the FDA) for years before you could ever treat a single patient with it, and you need to get reimbursed for the years and costs it took to develop the drug, the years and costs to get it through the proper regulatory authorities, and then the costs of producing and selling it. Without a patent, someone can just wait for you to pay for all the regulatory issues and then make it themselves after it's approved, all without paying a single dollar towards the costs of discovering it.

For other inventions, the costs of getting a patent (the USPTO's different fees, the cost of paying a patent agent or attorney to draft and work on the patent, the cost to you of making the invention a reality outside of your mind, and more) should be less than what you could get for it.

Patents are pursued in each country that has a patent system. If you pursue a patent in the US, your invention won't be protected in France unless you also pursue protection there. If this is your decision (i.e., it's your invention alone and no one else has the rights to it), then you have to decide if this is worthwhile. For example, 5 countries in Europe represent over 70% of the total European market (the UK, Germany, France, Spain, and Italy).[60] You may want to protect your invention in smaller markets as well,

[60] RegioData Research, "Purchasing Power Indices Europe - Edition 2012," 2012. www.regiodata.eu

but you will need to pay the costs for doing so *in each country*. Patents get expensive quickly.

Design patents

Design patents protect the look of something that is aesthetic and doesn't affect how the thing works, e.g., the special, distinctive signature curve of the handlebars of Bicycle Manufacturer X's bikes can be protected by a design patent because it does nothing for how the bike actually works. You are much more likely to see these if you are at a company (of any size, startups included) than if you are at a university.

Plant patents

Plant patents protect a new kind of plant that was made with human ingenuity that is not sexually reproduced (i.e., not from seed or from pollination). If you made it in a dish, it might be patentable. If it exists as a transgenic plant that you propagate using cuttings, it might be patentable. The length of time that these plants are protected is different, but otherwise getting a plant patent is similar to getting a utility patent.

Copyright

A copyright protects things like this book; it's not an invention and I can't protect the ideas contained in it, but I can protect how I wrote that idea down (or painted or drew or did a specific interpretive dance to communicate that idea). Plagiarizing a book is violating copyright. Making unauthorized copies of someone's drawing is violating copyright.[61]

---

[61] I have no idea how you'd sue someone for copyright infringement over your interpretive dance, but the defense would be great: "As you can see in the video, this sweaty fellow clearly stole my dance." "If that

Typically, to get a research paper published, you sign away your copyright to the publisher. If it is your book (e.g., you write a textbook on European insects called *The Battle of Swatterloo: Austrian Mosquitoes in History*), then you may be signing away copyright but getting royalties in return.

Trademarks

Trademarks are likely something your employer deals with, not you. The Coca-Cola™ logo (and name, as you can see) is trademarked. Your university's logo or mascot is likely trademarked. The Nike™ swoosh? Trademarked. Your favorite comic strip character? Trademarked. It is unlikely that you'll trademark anything in the course of your research, but an IP attorney can help you if you do; if you start a company, however, trademarks will be important for brand recognition.

Trade secrets

Trade secrets are things that are essential to a business that are not patented but kept secret. Referring to Coca-Cola™ again, the recipe is a trade secret; if you patent something, it becomes publicly accessible when the patent term is up, and therefore companies will keep certain things secret instead of patenting them. If you signed a non-disclosure agreement when you started the job, there are likely trade secrets that you will be arrested (or otherwise prosecuted) for revealing. Don't try to rationalize why *this* secret isn't stealing or an issue, just don't reveal it. These are typically only an issue once you get to a certain height in the business' hierarchy.

---

represents your dance, I'm not sure why you really want it."

Chapter 11

Grants and grant types

The Golden Rule: Whoever has the gold makes the rules. -The Wizard of Id[62]

Well, we've discussed saving money in your lab, and now we need to start discussing how to bring money in. This is not an extensive discussion of grant writing (which changes each year as the different grantors change their requirements, like word counts, page limits, and formatting) but is aimed at helping to focus your grant searching and helping you apply for grants more quickly than if you had to find all this out by yourself.

To start, if you're a graduate student, you should start looking for grants at the end of your first year in graduate school. Why, you might ask? It turns out that there are multiple grants that you can't apply for after your second year of graduate school. As such, if you start writing at the end of your first year (once you've gone through some introductory coursework and have some idea of what you're talking about) and apply within your second year, you can apply for the second-year grants (for example, the NSF Graduate Research Fellowship program only allows one application during your first or second year of graduate school). If you find out about these grants during your third year of graduate school, you're out of luck.

---

[62] *The Wizard of Id* by Johnny Hart and Brant Parker, 1967.

If you're a postdoc, you should start looking for grants very soon after you arrive. Ideally, you would start looking for grants once it becomes apparent that a) your boss has funding, b) you actually get along with your boss, and c) you will be supported for a second year (so that you won't run out of rent money in the meantime). It's not fair, but the lab's overall funding state can be used against your application. If you're serious about becoming a professor, a track record of funding will help you get noticed and help you get future funding (because we all like to support a winner).

If you're beginning faculty, you need grant money or else the lights get turned off. You are in the toughest boat of all, because other people are looking to you to have funding in order for them to continue to have a place to work, and getting grants is part of how you will get tenure. It's high-pressure and it will take up a lot of your time. Everyone should be a little kinder in the midst of the grant season.

Grant types and grantor types

It should be noted that there are many different types of grants for many different purposes:

- Travel grants, for travel to and from a conference or training opportunity;
- Equipment grants, for purchasing a specific piece of equipment or software;
- Research grants, which pay for all the associated costs of research, like salaries, equipment, consumables, etc.;
- Dissertation research grants, which are for improving or supporting someone's dissertation work, and just dissertation work;

- Tuition grants, many of which are specifically for underrepresented groups to join specific disciplines;
- ...and others.

These grants are split amongst different types of grantors, which you will likely need to become familiar with.

Federal/governmental grants

Governments provide the bulk of research funding for all types of grants. I have received a training grant and an equipment grant from federal funding. If you are seeking a research grant that will support an entire lab, equipment grants to buy expensive pieces of (perhaps shared?) equipment, or tuition grants, then you will need to look up the different governmental agencies (remember that most of them don't communicate with each other) that could support your research.

Nonprofit grants

Nonprofit grants include grants from, for example, the National Cancer Society or other nonprofit groups that work toward a specific cause. Many of these groups have grants that you can apply for, though for some nonprofits the amounts of money that they can afford to give make it sensible to use these as research grants for students and postdocs or as a supplemental source of income for a directed project. I received a grant from an honors society that allowed such applications.

I also found that my school, also a registered nonprofit, had a small dissertation improvement grant that I applied for and was awarded. If you are a student or postdoc, check with your school (and department, and program, and...) for any training grants or fellowships that they happen to administer. Check early; these are competitive. There are also new faculty grants administered by some of the same folks; if you don't ask, you won't learn about them until just after you could have applied. Ask early.

If you are involved in specific translational research, you have more options than if you are involved in basic research, though if your basic research is in an area of interest (e.g., cardiac biology, what with the modern world's increasing rates of heart disease) you also have a good chance when applying for these grants. In the worst case, they don't list a grant program, and you call and ask them if they have one. The worst they can tell you is no.

For-profits (companies)

Companies also provide research funding, though these come with more strings attached than research grants from the government or nonprofits. Company funding can be unobtrusive (for example, perhaps the area you are investigating is of particular interest to the company and they don't want the extra hassle of negotiating exclusive ownership of the data), but it can also get very complicated very quickly. It may require, for example, that you agree to share ownership of all the data you collect, or that you will give them ownership (via patent rights) of any invention that you make during the course of your research under that grant.

These are complicated and you will likely (unless you're a part of the company funding the research) need help from someone with experience in contract law. Your institution likely has such help and can work with you to make sure the agreement works for all parties. As research funding from governments grows more scarce, these grants (previously thought unacceptable by some researchers) will likely grow more acceptable and, frankly, necessary.

Research groups and consortia

Does your school have a student government? Is there a student government for graduate students? You can apply to such a group for travel funds or for a specific (small) research project. If your research is part of a consortium, the larger consortia also offer grant support (though not typically enough to keep the lights on, it can be good for travel to a conference).

Don't forget the group hosting the conference- they also typically offer travel grants, which you will likely need if you are unable to afford the travel otherwise. I have seen some travel grants require the awardees to help with the background work in the conference in some way. If you get such an award, do not complain! You are being given a free week-long internship where you can see what goes on to make a conference happen. You may, like I was, be lucky enough to help host your own conference (for my graduate program, in my case) and such experience will help your own gathering to be a huge success.

Which grants are available and when can I apply? An incomplete list

EU Grants

EU Commission on Research and Innovation

- Marie Sklodowska-Curie Actions: support PhD applicants OR postdoctoral applicants

- European Research Council: supports postdoctoral applicants
- Future and Emerging Technologies Actions: support research collaborations between science and engineering researchers with the intent of developing entirely new lines of technology

UK

- Erasmus Programme: undergraduate, graduate, and postdoctoral support
- Royal Society grants: support postdoctoral science research
- Euraxess UK: supports researchers' career development
- Commonwealth Scholarships: support students who hail from a country that was formerly a part of the British Empire ('Commonwealth countries')

US Government

- STEMGradStudents.science.gov offers multiple funding opportunities for work that meets a US government objective (multiple groups- the EPA, DOE, DOT, NSF, and others).
- NSF Graduate Fellowship: supports graduate student training in fields of interest to the US government (for up to 3 years), but you can only apply within your first two years of graduate school. Only open to US citizens (national or permanent resident status is also acceptable).

Graduate Students

- F30, Ruth L. Kirschstein Predoctoral NRSA for Dual-Degree Fellowships: supports students who are pursuing two degrees at once, e.g., MD/PhD students.
- F31, Ruth L. Kirschstein Predoctoral NRSA: supports students who are pursuing a PhD in a field that gets NIH funding, i.e., a health-related field. For example, neuroscience is supported but particle physics is not.
- F99 to K00, Predoctoral to Postdoctoral Transition Award: supports a student as they move from getting a PhD to being a postdoc.
- R36, Dissertation Award: pays for materials or supplies for dissertation research. Not renewable, i.e., you can only get this once.
- DP5, Early Independence Award: supports a researcher who moves directly from getting their PhD to leading a lab, i.e., no postdoctoral period.
- T32, Ruth L. Kirschstein Institutional NRSA: this grant is pursued by institutions (read: universities) to support select students on the grant.
- T35, Ruth L. Kirschstein Short-Term Institutional NRSA: this grant is pursued by institutions to support students over the summer
- T90/R90, Ruth L. Kirschstein Interdisciplinary Research Training Award (T90) and Combined Research Education Grant (R90): supports interdisciplinary training programs for undergraduate, graduate, and postdoctoral trainees

- D43, International Research Training Grant: supports research training programs to foster international research collaborations.

Postdoctoral and Early-Stage Researchers

- F05, International Research Fellowships: provides collaboration opportunities for foreign postdocs
- F32, Ruth L. Kirschstein Postdoctoral NRSA: supports students who are pursuing a PhD in a field that gets NIH funding, i.e., a health-related field. For example, neuroscience is supported but particle physics is not.
- K01, Mentored Research Scientist Career Development Award: provides support for a postdoc or early-stage researcher (staff scientist, research professor, tenure-track professor)
- K02: as K01, but for early- to mid-stage researchers (just a little later in their career)
- K07, Academic Career Development Award: for developing or enhancing educational materials (curricula, courses, etc.)
- K08, Mentored Clinical Scientist Research Career Development Award: for early-stage researchers in clinical disciplines
- K22, Career Transition Award: provides support to develop an new, independent research program (postdocs and early-stage researchers)
- K23: same as K22, but for clinicians who do research
- K25, Mentored Quantitative Research Career Development Award: as the other mentored research awards, but for people with backgrounds in 'quantitative' fields, like statistics and engineering, to do biomedical research

- K43, Emerging Global Leader Award: to support researchers in low- to middle-income countries with a faculty position and some research funding
- K76, Emerging Leaders Career Development Award: for supporting clinician-researchers
- K99/R00, Pathway to Independence Award: highly competitive award that supports a postdoc and continues to support them in a faculty position (at least initially)
- DP2, NIH Director's New Innovator Award: an award given for "highly innovative research projects" by early-stage investigators
- R25, Research Education Program: for developing or implementing an education program in specific areas
- T32, Ruth L. Kirschstein Institutional NRSA: this grant is pursued by institutions (read: universities) to support select students and postdocs on the grant.

Foundations and Nonprofits

Too many to list! Look up the foundation(s) in your field of interest to see what is available. In the life sciences, the American Heart Association, the Alzheimer's Association, the American Association of University Women, and the American Association for Cancer Research all provide some sort of fellowship opportunity (or multiple). I have seen one nonprofit foundation that shares equipment with researchers who study and develop lasers.

Note that nonprofit scholarships/fellowships include those offered by the specific university of interest. Many universities offer their own scholarships but, strangely, a number of them are simply unknown and therefore receive few applications.

Many nonprofit foundations have money and/or equipment to give for research but not a huge marketing/advertising budget. Go look for them!

For-Profits

Yep, everyone likes to get in on supporting research. Not every for-profit company has a scholarship, but there are some rare finds. For example, the Thermo Fisher Scientific Antibody Scholarship Program offers funding to undergraduate and graduate students (sorry, postdocs). Direct collaborations between industry and academia are increasing, and companies are often willing to foot part of the bill (as a grant) for research in their specific field of interest; however, these come with many more strings attached than a (much more vigorously contested and competed over) government or foundation grant. You will need to decide if the costs are worth the benefits (as I wrote earlier in this book, read the fine print).

Chapter 12

Nuts and bolts: grant writing tips

Cover letter and contact

If you have the opportunity to speak to a direct contact (aka training officer or a similar name) for the granting agency, do so. Ask what study section they suggest that you request in your cover letter. If the contact is willing, send them a copy of your specific aims to double-check that those aims fit the goals of the granting agency; yes, I know that *you* think they fit the agency's goals, or else you wouldn't be applying, but it's more important that the *agency* thinks your research fits its goals.

Regarding the cover letter, check the instructions for applying for that grant on how to write your cover letter after checking that you are going to the right place.

Formatting

Check and re-check the instructions. There are few things more maddening than working for months on a grant only to get rejected for formatting issues. Remember:

The Golden Rule: Whoever has the gold makes the rules.

You're asking for their gold, so you have to play by their rules. The formatting can change between years and grant cycles (e.g., from 10 pages to only 5 pages, 11-point font to 12-point font, etc.). The main thing to remember is that effective writing is concise writing: get your point across quickly and, as William Strunk said, "Omit needless words!"[63]

Preliminary data

There isn't necessarily going to be a specific section for your preliminary data, but you had better believe that you're going to need it anyway. This can be put into a preliminary data section, the background section, or under the research approach as you describe the experiments you want to do under that grant.

As grants continue to grow more competitive, having additional preliminary data (that are relevant) will help your application to be competitive. That said, I do understand the chicken-or-egg scenario ("If I had completed the study already, I wouldn't need to ask you for money!").

Background

You need to include enough background that the reviewers will understand why you developed your specific research hypothesis, and no more background than that. This is where proofreaders (particularly those that don't have much to do with the project) come in handy- you need someone who will read just what you've provided and determine if that is enough to understand why you have this hypothesis.

The goal is to provide enough background to facilitate understanding while leaving yourself enough room for all the other things that have to go into the application, e.g., did you describe the facilities available to you? Personnel? How will the

---

[63] Strunk and White, "The Elements of Style." 1959.

information be disseminated to the public? There are many things that funders may require you to answer in your application that have nothing to do with the science.

The background section will include your specific aims, which will be further developed in the research approach section. If your specific aims are given space of their own, then the following is a possible format you may use:

- Introductory paragraph with an interest-grabbing first sentence
  - What is the current level of knowledge?
  - What are the gaps in the current knowledge?
- What and why paragraph- what are the long-term goals of this research and why are they important?
  - Make your central hypothesis very clear and justify it!
  - What is the objective of this specific application?
  - Why will we be better off if you get to research this thing?
- The aims- list separately, 2 to 3 at most, and make sure that they are all related to the central hypothesis
  - Must be hypothesis-driven
  - Don't make them depend on one result, e.g., if A doesn't work, and then there is no reason to try experiments B, C, D, and E; if your plan is like that, then that's not a well-designed plan that is likely to get funding
- Expected outcomes and overall gains

Research approach

The research approach section will include the following:

1. Figures
2. What you are doing (methods)
3. What you expect (expected results)
4. What might go wrong (possible issues or alternate outcomes)
5. What you plan on doing if your initial experiments don't work

You need to include a plan for what you will do if your experiments don't work. Again, if your grant application hinges on experiment 1 going correctly, then it won't get funded because if experiment 1 doesn't work and aims 2-4 all go into the trash can, it's not worthwhile for the agency to support the entire grant.

Make this section systematic and easy to follow; A logically develops into B, which is logically followed by C, for example. As one possible suggestion for a specific aim/approach format:

Aim 1: Title
Rationale: why does it make sense to ask this?
Design: What are my independent and dependent variables? What am I actually testing and what are my controls?
Expected results and interpretations
Potential pitfalls and interpretations

Significance and innovation

The significance section is crucial and will explain the following:
1. What are the current barriers in our understanding of this field, and how will your work address them?

2. How will your research improve knowledge in the field as a whole, interpreted broadly (e.g., I am working in this brain injury model, which will help us understand brain injury, which will help us understand how the brain works, which may help us understand how cells throughout biology work)?
3. How will the field change if your goals (specific aims) are achieved?
4. If the application is for a training grant or includes a training component, how will this be significant for your career as a scientist?

The innovation section (if applicable) will explain the following:
1. What new theories are being tested in this research?
2. What new methods or approaches are being developed in this research?
3. What new therapeutic approaches or possible medical interventions are being developed in this research?
4. What new thing is being developed in this work that this granting agency cares about? For example, don't send your solar panel project to the Bovine Appreciation Society (if they suddenly start supporting research through grants).

If there is no specific section to mention the significance (importance) of this research or how it will innovate and change the field of research, you will need to put this in the application somehow. If the reviewer can't figure out why your work is important, they won't fund it. If you can make it really easy for the reviewer to understand why what you are doing is important, it's easier for them to fund your work.

If you can't come up with a good reason or three to fund your work, you might need a new research question (or approach). If you think you're being too condescending or dumbing things down too much, you're probably not, and you can keep it. If you're really concerned, ask someone unrelated to your project to give you honest feedback.

Remember the fact that when you are editing your own work, you forget what it was like to just enter your field and lack all the knowledge about the field that you now possess. Because of this effect, you can write in jargon all day and understand it perfectly well, whereas your reader might not!

Project summary and abstract

Remember the following when you write the project summary and abstract:

1. This part is going to become public knowledge (or at least publicly accessible).
2. This is the first thing that the reviewer reads and is the basis for which review group your grant is assigned to.

The easiest way to develop this (if you've written every other part of the grant first) is to take 1 or 2 sentences from your background section on the significance of the problem and put them in. Afterward, clearly state or paraphrase your hypothesis and then state the specific aims you have for this grant.

Stay within the word and page limits set by the granting agency.

A related document (that may or may not be required) is a public relevance statement (may also be called a project narrative)- this states why the public would find this work important and should be written for a non-scientific audience.

Facilities, etc.

This describes the facilities that are available to you, if the grant requires it. This can include the laboratory space, your personal office or computer, clinical space (if applicable), animal facilities and personnel, core facilities and shared equipment, and fellow researchers with specific expertise that you can ask for help.

You may also be asked about biohazards, biosafety compliance, animal use protocols, why the sponsor and your institution are great choices for helping you perform this research, and your respective contribution to the research (or writing), among other topics.

Reference letters

We never do quite get away from these. If you are applying for a training grant, your mentor needs to write a section about how you are just magical and rainbows spring from your every step...or at least why you're such a great researcher (you could also work for a really famous person and hope that their fame rubs off on you and stains your clothes). If you're the mentor, you get to write this section and describe the support available, previous trainees and where they are now, the training environment, and other questions.

If you are early enough in your career, you may be asked for reference letters to indicate why you are a safe bet for this granting agency to hand over $500,000 to you and why they can expect that you'll actually do research with it instead of run off to a tropical vacation. Choose people who will answer honestly about you.

Resubmission

If you do not get approved but are given the opportunity to resubmit, make sure that you respond to and answer every concern or question given by the reviewers. If you are not seen as responsive to their concerns, the reviewers can easily decide to stop reviewing your application. The reviewers are moving through a large pile of grant applications, and you don't want to make it easy for them to choose to throw your application out in the trash.

Epilogue

My sincere hope is that you are now empowered to survive grad school (or your new job) and run a lab effectively, learning from my mistakes (or wisdom).

Thanks for reading! If you liked this book, please leave a review on Amazon, iBooks, or Goodreads!

Good luck and happy (scientific) hunting!

Glossary

Numerical

R01: This is a National Institutes of Health (NIH) research grant and pays for an entire research project: staff, facilities, supplies, the works.

T32: This is a National Institutes of Health (NIH) training grant that pays for a student's graduate training: money to live on, tuition, fees, and perhaps some supplies.

A

Adjuncting: teaching classes as an adjunct professor.

Adjunct professor: a teacher who is hired on a per-class and per-semester basis. This teacher has no benefits or long-term relationship with the school as far as employment is concerned. Typically does not make enough money to raise a family (or often themselves) by adjuncting alone; many need to teach multiple classes per semester (3-7) to make enough money to live on. A source of once-a-year angst in the news cycle as the media rediscovers that these folks exist.

Alternative career: historically, any career for someone with a PhD that was not a professorship was referred to as an alternative career. Even though professorships are becoming more scarce, the term remains.

Alumni network: each university keeps a list of the people who graduated from that school. Sometimes these alumni meet on their own and try to help other graduates of that school in some way (networking, donations, etc.).

Audit: an official inspection of your accounts by the government or some other authorized group. If there is a suspicion that you've been lying about your income or taxes, you would get one of these.

Automagically: portmanteau of 'automatically' and 'magically.' Meaning something that happens on its own but seemingly out of thin air or for no good reason.

B

Back-of-the-napkin math: an inexact estimate or the practice of using rounded-up or rounded-down numbers to make your calculations easier.

Budget: a list or record of what you spend money on in a given month. Can also be your plan for how your money will be spent, saved, or given away in a given month.

C

Career center: a place at your school or university that has career counselors and other things available to help you find a job after school.

Cheap: miserly, stingy, unwilling to pay more than the absolute minimum for something. Can also describe an item of questionable quality.

Collaboration: a sometimes formalized agreement where you will work with another laboratory or person on a project of mutual interest. Often has defined roles for each person, e.g., Dave will do the western blotting, Eliza will do the mass spectrometry, and Frank will do the statistical analysis.

Copyright: legal rights regarding certain artistic creations (like this book you're reading) that grant exclusive rights to the author or copyright holder. For example, I can't copy the latest Stephen King novel and start selling it myself; the copyright holder can (and will) sue me for the money and damages.

D

Demo equipment: research equipment that is used, but was only used to show others how to use the product or was simply put out at trade shows for people to look at.

DIY: Do It Yourself. Build it yourself, maintain it yourself, do the work so that you don't have to pay someone else for it.

E

Entrepreneur: someone who starts and/or organizes a business or businesses, which is much more risky (and potentially rewarding) than being an employee.

F

Faculty: a term to describe a professor (faculty member) or all the professors as a whole (the faculty).

FAFSA: The Free Application for Federal Student Aid. If you want some sort of government assistance for school, you will need to fill out the FAFSA. This can also be done at the graduate level.

Filing taxes: the process of settling your accounts with your government (state, federal, local, etc.) regarding your taxes. May result in a tax refund (a check for you) or additional taxes for you to pay.

Fine print: slang term for the additional terms of a contract that are always printed in very small font so as to a) fit much more onto a page, b) make it hard to read, and therefore c) make it much more likely that you will just agree to the terms.

For-profit: a typical company that is organized to sell goods and/or services for a profit. Typically is privately owned or has shareholders. This term is mostly used as a point of contrast to nonprofit groups.

Frugal: in this book, someone who spends money where it is necessary and does not spend it when it is unnecessary.

G

Garage sale value: something that is considered to have garage sale value is something that is worth only what you can haggle for it ($5 or less).

Grad school: a professional school that is differentiated from other professional schools in that you are a) pursuing a PhD and b) conducting research as a primary focus.

Grant: money that is designated for a specific purpose but you don't have to pay it back (unless they have some sort of language in the contract that states you must fulfill certain requirements to avoid paying it back). In general, it's free money that you must use for that specific purpose. Example purposes: travel to a conference, paying for research materials, paying someone's salary, or setting up a new service facility for the tri-state area.

H

Hypothesis: in research, your proposed explanation for why something happens, which you go investigate to determine if that hypothesis can be disproven.

I

Impostor syndrome: that feeling that you shouldn't be in grad school because you don't belong, that everyone else gets it and you do not, and that you are doomed to fail anyway. It's not true, but it is felt by the majority of graduate students.

Insurance: a contract with an insurance company that defines when and how you would be paid if something happens that is covered by the contract. For example, a term life insurance contract means that the insurance company pays money to your surviving family/designees if you die during the term (which is likely quite unexpected), flood insurance means they pay money to you in the unlikely event that your house floods, and disability insurance means that they pay money to you in the event that you are permanently disabled. If the thing in question never happens while you hold the insurance contract, the insurance company doesn't have to pay you anything but you paid them for the privilege of holding the contract.

Intellectual property: property that was generated from someone's mind, e.g., written works, musical compositions, inventions, cartoon drawings, and specific logos.

Ivy League: initially used to describe a sports conference, this now describes a group of exclusive schools that are believed to provide a superior education and can be shown to provide superior contacts if one wishes to enter banking, consulting, or other lucrative, competitive industries.

J

Jackpot: gambling term. A cash prize that accumulates until someone wins, e.g., the pool of money to be won in a lottery.

Job fair: a gathering of employers who meet in one place to meet potential recruiters and new employees. If you wanted a job, you used to need to go to these.

L

Lab: short for laboratory. If you're not spending a lot of time here, you soon will be (out of necessity).

Licensing: the process of allowing others to use your patented material/process in exchange for money, stock, or some other form of compensation. This allows the use of the patented material without getting sued.

Life insurance: insurance that pays out upon the death of the person who entered the contract.

Life science: branches of science that deal with living creatures, e.g., biology, biochemistry, botany, microbiology, etc.

M

Murphy's Law: That which can go wrong, will go wrong.

N

Negative control: in research, a negative control is something that should not show what you're looking for, e.g., a tissue that does not express the protein you're looking for, an alloy that does not bend at 200°C when the alloy you're examining should, insulation that does not form a complete circuit when your insulation should, etc.

Networking events: social events where you are supposed to go exclusively for the opportunity to meet people for the sake of your career.

NIH: the National Institutes of Health. Primary grant funder of life science grants in the US.

Nonprofit: a company or organization that is not organized to pursue profits. Typically, these organizations lack shareholders and operate under different laws than for-profit companies. Examples of nonprofits: hospitals, universities, and foundations.

P

Patent: a monopoly granted by the government on the ability to produce, perform, or use an invention.

Penny wise and pound foolish: being economical or frugal with small amounts of money while spending recklessly with large amounts.

PhD: Doctor of Philosophy. The terminal degree for graduate students.

Positive control: in research, a control that should show the same phenomenon that you are looking for, e.g., a tissue that expresses the same protein you're looking at in another tissue, another fungus that responds to light like your studied fungus does, etc.

Postdoc: slang term for postdoctoral scholar, fellow, etc. Simply a shorter way of saying a scientist who got their PhD but isn't recognized as a full staff member somewhere.

Pre-med: Someone whose undergraduate education was geared toward applying to medical school to pursue a career in medicine.

Professional school: some type of schooling that is required after getting a bachelor's degree to work in a given industry. Examples include business school, physician's assistant school, dental school, medical school, and graduate school.

R

Recruiter: someone who is paid to bring people into an organization, whether that's a company, nonprofit, or governmental organization. They are not paid to help you achieve your dreams, they are paid to fill a position for an organization.

Reference letter: a letter written on your behalf by a colleague or senior person in order to tell the funding committee (or the person hiring you) that you're great, you're amazing, and the funding group can totally trust you with $500,000 per year.

Refurbished equipment: equipment that has been fixed up or checked that it still works to factory specifications. Worn-out parts should generally be replaced or fixed for something to be considered refurbished. Cheaper than a new model because of the wear and tear.

S

Sabbatical: paid leave for a professor that is typically used for traveling, learning a new skill, or further developing their career.

Salary: generally, getting paid once a month at a fixed rate; if you get paid a fixed amount no matter how many hours you work, you're salaried (receiving a salary).

Seat time: slang for time spent physically in a seat at a given location. Often equated by people with productivity because seat time is easier to measure.

Social media: a way to keep in touch with people that doesn't involve actually speaking to them. Also, significant use of social media is linked to depression and it's an effective distraction from all the other stuff you should be doing.

STEM: Science, Technology, Engineering, and Mathematics.

Stipend: money to live on while in graduate school. Because it's not paid employment, they have to have a different name for the money you are given while working as a student or on a grant.

Student loan: money that a student borrowed to pay the costs of going to school for a degree. Cannot be gotten rid of in a bankruptcy; this awful debt follows a person wherever they go.

Supervisor: the person who decides if you should be fired or not. If you're in grad school, this person ostensibly mentors you as well.

Syllabus: considered a contract between the teacher and their students for a given class. It is important to lay out all of your expectations in here so that there is written evidence that you said or intended something.

T

TANSTAAFL: There Ain't No Such Thing As A Free Lunch. Nothing is truly free; it may cost you your time, your patience, your dignity, or your money.

Tenure: guaranteed permanent employment unless you egregiously break the law, e.g., parking tickets won't take away tenure but being found guilty of murder or molestation will (or should). The Holy Grail of academia.

Tenure-track professorship: a professorship career track that has tenure as the reward after 7-10 years. This can be contrasted with research professorships, which are not on the tenure track.

Trademark: a phrase, word, or symbol that represents a company or group. These must be legally registered to be considered enforceable.

Trade secret: a company's secret. Anything patented must be revealed and becomes open to the public when the patent expires; as such, to truly keep something a secret forever, it must be a trade secret and is guarded much more strictly than a patent.

U

US: United States. Many of these different retirement plans are available only in the US and are due to the labyrinth we call the US Tax Code.

W

Wear and tear: slang for the typical wear, damage, and aging of vehicles, equipment, and large machines. If referred to as a cost, means the maintenance costs to defray the effects of typical wear, damage, and aging.

About the author

John Corthell earned a PhD in neuroscience from Florida State University. He left bench science due to an allergy but since then has pursued opportunities related to business development, lab management, management consulting, technology transfer, and patent law. He misses the lab bench but enjoys helping scientists through advice, analysis, and mentorship. He is still an avid reader, kayaker, cook, janitor, consultant, and explorer. He tells and greatly enjoys puns and other dad jokes (according to friends, he's practiced them since the 10th grade). He lives with his lovely family in the Midwest.

www.ingramcontent.com/pod-product-compliance
Lightning Source LLC
Chambersburg PA
CBHW020038040426
42331CB00030B/43